Lecture Notes in Computer Science 705

Edited by G. Goos and J. Hartmanis

Advisory Board: W. Brauer   D. Gries   J. Stoer

Herbert Grünbacher
Reiner W. Hartenstein (Eds.)

Field-Programmable Gate Arrays:

Architectures and Tools for Rapid
Prototyping

Second International Workshop
on Field-Programmable Logic and Applications
Vienna, Austria, August 31 - September 2, 1992
Selected Papers

Springer-Verlag

Berlin Heidelberg New York
London Paris Tokyo
Hong Kong Barcelona
Budapest

Series Editors

Gerhard Goos
Universität Karlsruhe
Postfach 69 80
Vincenz-Priessnitz-Straße 1
D-76131 Karlsruhe, Germany

Juris Hartmanis
Cornell University
Department of Computer Science
4130 Upson Hall
Ithaca, NY 14853, USA

Volume Editors

Herbert Grünbacher
Institut für Technische Informatik, Technische Universität Wien
Treitlstr. 3/182.2, A-1040 Vienna, Austria

Reiner W. Hartenstein
Fachbereich Informatik, Universität Kaiserslautern
Postfach 30 49, D-67653 Kaiserslautern, Germany

CR Subject Classification (1991): B.6-7

ISBN 3-540-57091-8 Springer-Verlag Berlin Heidelberg New York
ISBN 0-387-57091-8 Springer-Verlag New York Berlin Heidelberg

© Springer-Verlag Berlin Heidelberg 1993
Printed in Germany

Typesetting: Camera ready by author
45/3140-543210 - Printed on acid-free paper

Preface

This book contains papers first presented at the 2nd International Workshop on Field-Programmable Logic and Applications (FPL'92), held in Vienna, Austria, from August 31 to September 2, 1992.

The growing importance of field-programmable devices, especially of field-programmable gate arrays, was demonstrated by the increased number of papers submitted in 1992. For the workshop in Vienna 70 papers were submitted. It was pleasing to see the high quality of these papers and their international character with contributions from more than 20 countries. The following list shows the distribution of origins of the papers submitted to FPL'92 (some papers were written by an international team):

Australia:	1	Austria:	4
Canada:	1	Czechoslowakia:	1
Finland:	3	France:	2
Germany:	16	Italy:	3
Japan:	3	Norway:	3
Russia:	1	Spain:	4
Sweden:	4	Switzerland:	1
United Kingdom:	4	USA (incl. Hawaii):	21

From the 70 submitted papers, 23 were selected for this book. The first three papers discuss strategic issues and give surveys. Three papers deal with new FPGA architectures and five papers introduce methods for tools. The last twelve papers report applications focusing on rapid prototyping or new FPGA-based computer architectures.

We would like to thank the members of the technical program committee for reviewing the papers submitted to the workshop. Our thanks go also to the authors who wrote the extended papers for this issue and the reviewers for their timely work on all manuscripts. Especially we would like to thank Helmut Reinig for managing the reviewing process and handling the manuscripts and program planning.

Thanks to the sponsors of the workshop: Universität Kaiserslautern, Technische Universität Wien, IFIP Working Groups 10.2 and 10.5, Wirtschaftsförderungsinstitut der Bundeswirtschaftskammer, Wien and Bundesministerium für Wissenschaft und Forschung, Wien. We also gratefully acknowledge all the work done at Springer-Verlag in publishing this book.

June 1993

Herbert Grünbacher,
Reiner Hartenstein

Program Committee

Organizing Committee

Table of Contents

Overview of Complex Array-Based PLDs

Günter Biehl

ISDATA GmbH - Daimlerstr. 51 - W 7500 Karlsruhe 21 - Germany

Abstract. The first PLDs based on sum-of-products arrays were PLAs and PALs. Both are special cases of the more general PML, a fed back NAND array. Some CPLDs take up the idea of the PML for product term expansion, others use various methods of product term allocation. The multiple array architecture is the way to increase the pin count of PLDs. Their logic design requires to partition the logic in consideration of the interconnect matrix of the PLD. Limited interconnect is the reason for an additional placement problem.

1 Types of Programmable Logic

Field programmable logic devices can be divided into three main groups:

- *Ramdomly addressable memories:* PROM, EPROM, EEPRPOM, RAM
- *Array logic:* e.g. PLD, PAL, GAL, PLM, FPLA, EPLD, multiple array PLD;
- *Programmable gate arrays:* e.g. ACTEL, XILINX, Quicklogic.

Implementation of logic in memories is not a problem, provided that the number of address pins of the memory is sufficient. Logic design for programmable gate arrays is similar to synthesis for gate arrays. This paper concentrates on features of array logic.

2 Programmable Array Logic

2.1 PLA and PAL

The ancestor of array logic is the very flexible *PLA* structure illustrated in figure 1a). It consists of a programmable AND-array and a programmable OR-array. It allows the allocation of any number of product terms of the AND-array for each output function. The first field programmable devices on the market were PLAs, offered by Signetics and Intersil in 1975.

Real mass applications of programmable array logic was achieved by Monolithics Memories (MMI) 1978 by a restriction of the general structure: In the *PAL* (later GAL, PLD) structure illustrated in figure 1b) the product terms are allocated to the output functions according to a fixed scheme. Therefore the OR array is replaced by a much simpler hard wiring.

In a full PLA, common product terms of different output functions need to be implemented only once and may be shared by the individual functions. In a PAL structure each function must implement it's own copy of the common product term. For the implementation of a design of a certain logic complexity, a PAL therefore generally needs more product terms than a PLA.

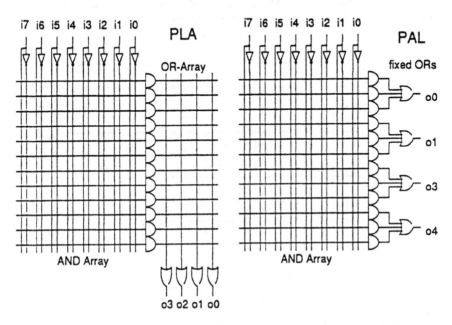

Figure 1: a) PLA structure, b) PAL structure

PLA and PAL enable the two level implementation of any boolean function of the inputs in sum of products form (AND/OR form), provided that the device has enough product terms.

Example 1:   y is function of 5 inputs a, b, c, d und e:
$$y = a\,b\,d + a/b\,e + b/c\,d + /b/c\,e + b\,d\,f + /b\,e\,f$$
This sum of products form consists of six product terms.

2.2 PML and ERASIC

Even more general than the PLA structure is the PML (programmable macro logic), which was indroduced by Signetics years ago (figure 2). Instead of a separated AND and OR array, the PML has one homogeneous NAND array. A small fraction of NAND outputs feed the output pins, most of the NANDs are fed back to the array. Thus the PML consists of very wide NAND gates, which may be arbitrarily connected via the feedback path.This structure enables NAND networks of any number of logic levels. The PML contains the AND/OR structure of the PLA as a special case: As we know from boolean algebra, a two level NAND/NAND network is equivalent to a AND/OR network.

Therefore PMLs offer the highest degree of freedom for logic synthesis, but unfortunately the highest degree of difficulty for a synthesis software too. Logic synthesis for PMLs is more related to multi level synthesis for gate logic than to PLD design. Unlike gate synthesis, the sole cost criterion for PMLs is the number of NANDs, as the cost of a NAND does not depend on it's number of inputs.

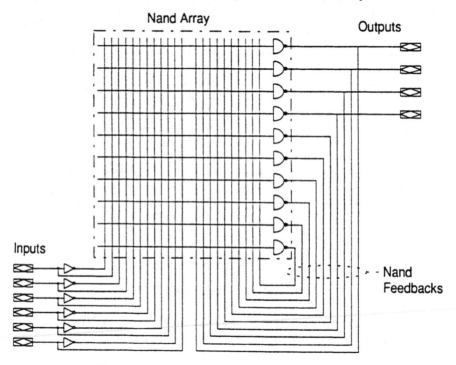

Figure 2: PML as the most general logic array structure

Example 2: the function y of example 1 can be transformed into a representation consisting of four NAND terms instead of six product terms:

$$y = a\,b\,d + a\,/b\,e + b\,/c\,d + /b\,/c\,e + b\,d\,f + /b\,e\,f$$

1. by factoring out 'a' out of the first two terms
$$y = a\,(b\,d + /b\,e) + b\,/c\,d + /b\,/c\,e + b\,d\,f + /b\,e\,f$$

2. by factoring out '/c' out of terms 2 and 3
$$y = a\,(b\,d + /b\,e) + /c\,(b\,d + /b\,e) + b\,d\,f + /b\,e\,f$$

3. by factoring out 'f' out of terms 3 and 4

$$y = a (b d + /b e) + /c (b d + /b e) + f (b d + /b e)$$

4. by factoring out '(b d + /b e)' out of the three terms
$$y = (b d + /b e) (a + /c + f)$$

5. introducing the intermediate variable $z = (a + /c + f)$:
$$y = (b d + /b e) z$$
$$y = (b d z + /b e z)$$

6. transformed into NAND:
$$z = /(/a c /f)$$
$$y = /(/(b d z) /(/b e z))$$

Hence the implementation of y (and z) in the PML requires only 4 NAND terms according to figure 3, whereas a PLA or PAL implementation requires 6 product terms. Notice that the NAND form has increased delay because it requires three levels of logic.

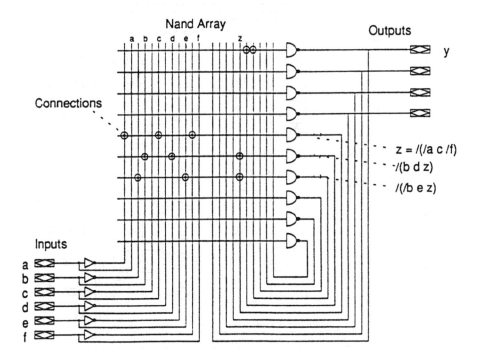

Figure 3: NAND implementation of function y: only 4 NANDs instead of 6 product terms

Up to now no vendor independent PLD compiler offers an optimizing NAND synthesis for PMLs. This might be the reason, why the ideal PML stucture is so rarely used today.

The ERASIC devices of EXEL are closely related to the PML. Instead of the NAND array the ERASICs consist of a fed back NOR array. In the same way as we can implement a two level sum-of-products in NANDs, we can implement a two level product-of-sums in NORs. The real benefit of the ERASIC's NOR array can only be made available by a multi level logic synthesis. The lack of powerful synthesis software might be main reason for the ERASIC's unsuccessfulness.

3 Increasing the Pin Count by Multiple Array Devices

Today PLD designers require PLDs with up to 100 - 200 I/O pins. When trying to increase the complexity of PLDs without sacrificing speed, PLD manufacturers obviously cannot increase the arrays homogeneously. The number of product terms is also limited by the speed required for most PLD applications.

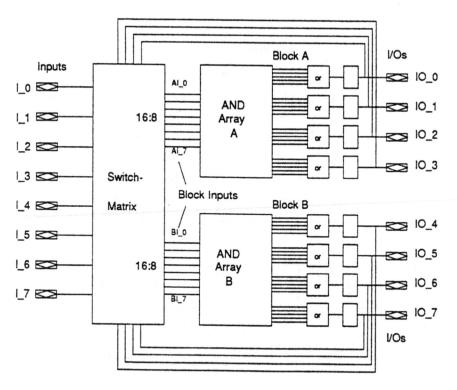

Figure 4: hypothetical example of a multiple-array-PLD consisting of two PLD blocks

Therefore all the new architectures of array based PLDs make compromises. Instead of a big homogeneous logic array, which would be most desirable from the designers point of view, they contain 2 - 32 smaller PLD blocks, which can be connected via a programmable interconnect array (or switch matrix). Figure 4 illustrates the principle using a hypothetical device which consists of the two PLD blocks A and B. The eight inputs of each block can be fed by a selection of the total 16 signal sources of the device (8 inputs and 4 feedbacks from each block).

Even if todays multiple array PLDs have much more than 100 signal sources (inputs and feedbacks) the individual arrays have only 20 - 40 inputs. Thus very large devices can be build which operate at high speed. The additional delay which comes from the interconnect array is moderate and does not depend on the particular signal path. Because of this regular structure the delay is easily predictable.

The details of the switch matrix depend on the particular device type. The best case is switch matrices which are able to route any selection of signal sources to any PLD block. For this flexible type of switch matrix the design method may be rather simple: Equations which share input variables are preferrably grouped together in one block, as long as the total number of block inputs is not exceeded. This problem is known as block partitioning.

On the other side there are switch matrix architectures which have strong impact on the logic design process. If the switch matrix consists only of small multiplexers, only few combinations of signal sources can be routed to a PAL block. In this case the good block partitioning is not enough. Moreover it is necessary to pin the signals such that the required feedback paths are available in the interconnect array.

Examples of multiple array PLDs are: the MAX families of Altera, the MACH family of AMD, Xilinx (PlusLogic) Hyper-family, and the pLSI-family of Lattice.

4 Architectural Tricks to Increase Logic Complexity

Very often not only the pin count, but also the product term count available per output, is a bottleneck of PLD devices. Therefore all the different architectures of newer complex PLDs offer very different compromises to allow logic complexity while maintaining high speed.

4.1 Increasing the number of product terms for some outputs

The simplest method makes use of the fact, that almost never do all equations of a PLD have the same complexity. Therfore the number of product terms varies for different outputs of a PLD. Thus the total number of product terms of a device is moderate and nevertheless some few functions may be very complex. Certainly this feature is one of the main reasons for the sucess of AMD's 22V10 architecture with 8 to 16 product terms per output.

Newer device architectures avoid delay problems which come from large numbers of product terms. They contain only few (3 - 7) private product terms per macrocell, but there are different expansion mechanismes:

One of those is used in AMD's MACH family and Lattice's pLSI family. Their *product-term-allocation* enables a macrocell to allocate the product terms of its neighbours if needed. A first approach of this method was was used years ago in MMI's product-term-sharing PAL series 20RS10.

Figure 5 shows as an example the product term allocation array of the pLSI 1032. The programmable 4x4 array enables an arbitrary allocation of the 4 product term groups to the outputs of the logic block.

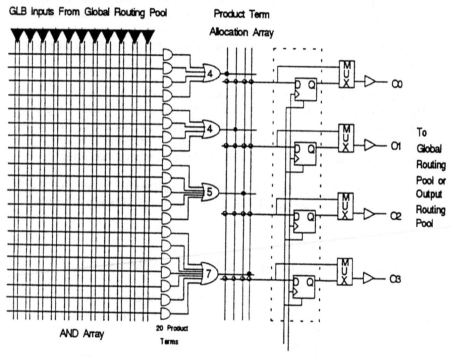

Figure 5: Example for Product-Term-Allocation: pLSI 1032

One can configure

- one function of 20 product terms or
- two functions of 12 and 8 product terms or
- three functions of 9, 7, and 4 product terms or
- four functions of 7, 5, 4, and 4 product terms

in one block.

The product term allocator of AMD's MACH family is very similar: Four product terms are available per macrocell. Each macrocell can borrow the product terms of up to three of it's neighbours and thus functions may use up to 16 product terms.

4.2 Multi level logic

In Altera's MAX family, each function has even only three dedicated product terms. At the first glance this seems to be few, but it is compensated by a very clever expansion mechanism. As illustrated in figure 6, each macrocell brings in two expander terms, which can be allocated by any other macrocell of the PLD block. The logic type of all the terms is NAND.

The three private NAND terms are tied to a NAND in the macrocell. This NAND/NAND architecture is equivalent to a AND/OR form as known from Boolean algebra. Thus this part of the MAX device implements the sum-of-products common to all PALs.

The expander terms are directly fed back to the NAND array, just as the NAND feedbacks of Signetics' PML family. All the terms of the PMLs can be considered as expander terms. Thus the expander terms enable the implementation of multi level logic as described above for the PMLs. Furthermore, by cross cuppling of NAND terms one can build asychronous filpflops.

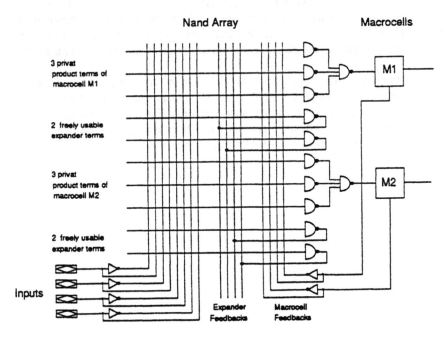

Figure 6: Product terms and expanders of the MAX macrocell

Unlike other PLDs, when designing logic for MAX devices, the product term count of the sum-of-products form is not sufficent to decide wether a macrocell can implement a function or not. Each function of more than three product terms can be implemented by use of the expander terms. In this case the objective of an optimization procedure is the use of as few expanders as possible, as the total number of available expanders is limited.

The drawback of the expanders is increased delay which is dependent on the levels of expanders. A second objective of an optimization procedure is to keep the level of logic as low as possible. Both criteria are considered in the MAX fitter of ISDATA's LOG/iC Compiler.

4.3 Partitioning of the product term array in the time domain

A completely new aproach speeding up sequential devices was introduced by National Semiconductor. The MAPL family of NS restricts the general PLA structure. The product terms of the AND-array are not allocated to the functions as in PALs, but are allocated to the states of a FSM. The consequences of this partitioning are as follows.

The product term blocks of the MAPL as shown in figure 7 must not be confused with the PAL blocks of multiple array devices. For the MAPL, in each clock cycle of the FSM only one single block of product terms is active. Of course this does not reduce the chip area of the whole array, but it reduces the power consumption of the device. The product terms of the current block are enabled by decoding of the select bits. It is the contents of the select register which decide on the selection of the block.

By an appropriate state asignment of the FSM and by the corresponding placement of the product terms within the AND array one can activate, at any time, only the product terms necessary for the actual state transition. Even if, at any time, only one block is active, functions of more product terms than available in one block can be built. Of course this architecture is for sequential logic only.

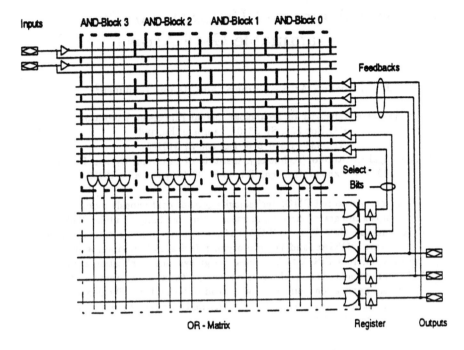

Figure 7: Architecture of the MAPL (simpified)

Extensions of the MAPL devices are available which combine the AND blocks with a classical PAL block, thus enabling combinatorial logic too.

5 Conclusion

The ideal architecture of array based PLDs was a huge NAND array (or NOR) of 100-200 inputs and at least as many fed back NANDs. Because this array is not feasible, compromises in the array size and architecture are necessary in order to achieve the speed requirements. The PLD vendors offer solutions based on very different compromises. The selection of the best solution depends on the application and can not be decided generally.

Technologies and Utilization of Field Programmable Gate Arrays

Jouni Isoaho[1], Arto Nummela[1] and Hannu Tenhunen[2]

[1] Tampere University of Technology, Signal Processing Laboratory
P.O.Box 553, SF-33101 Tampere, Finland
Tel. 358-31-3161876, Fax 358-31-3161857
[2] The Royal Institute of Technology, Institute of Electronic System Design
S-10044 Stockholm, Sweden. Tel. 46-8-7907810, Fax 46-8-103925

Abstract. Advanced CMOS technologies provide continuously more dense and complex integrated circuits, increasing the possibility to utilize Field Programmable Gate Arrays (FPGAs) as a competitor of Mask Programmable Gate Arrays (MPGAs) and as a prototyping device. In this paper, we will overview different FPGA technologies, design systems, application areas, and future trends. The emphasis is on looking these issues from the designer and application point of view instead of technology. The denotion FPGA is used to cover both the FPGAs and complex Programmable Logic Devices (PLDs).

1 Introduction

Traditionally, MPGAs are used for rapid prototyping of ASICs to speed up considerably the design process compared to full-custom design [20]. As the utilization of low power CMOS technologies made it possible to integrate large amounts of logic into a single chip, the programmable devices became a reasonable alternative to MPGAs in several applications, like in consumer electronics, industrial control systems and even in communication. In late 1980s, the size of FPGAs reached several thousands of usable gate array gates making them a reasonable solution for small size and volume ASICs. The development also started the boom of rapid prototyping of ASICs. Due to the fast technological and architectural development of FPGAs, the economical break-even point between MPGA and FPGA for the project is not simply anymore. The selection of the prototyping approach also depends on the application, the stability of the specification and the type of prototyping needed.

Because there are no standards for comparing the different types of FPGAs, and a wide diversity of FPGA technologies [3] and development tools are available, the selection of a suitable FPGA solution (FPGA technology and tools) for the application is not straightforward and general rules cannot be provided. To facilitate the selection and to form good quidelines for it, updated technological information of devices and supporting design tools are needed. Also the trends of future development should be known in order to create the long range strategy for the design group.

---

[3] over 30 vendors if both large and small devices is taken into account

2   Technology Overview

The comparison of the technology development of memories, MPGAs and FPGAs is presented in Figure 1 [18] [1] [12]. Memories, as a state of the art application, is used to present the technological limits from the beginning of FPGA history. Today, FPGAs and MPGAs are processed using the same size of dimensions and they are all the time coming closer to the design rules used in memory devices. In the price competition, it has to be remembered that FPGAs are high volume standard products.

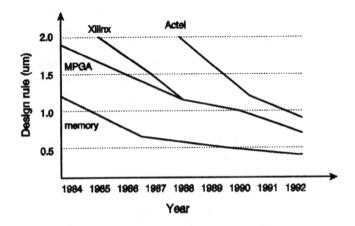

Fig. 1. Smallest feature sizes of memories, MPGAs and FPGAs.

As compared to MPGAs, FPGAs provide faster inhouse prototyping (implementation) with smaller NRE costs, but at the same time, they have lower operating speed and logic density on silicon. System speeds up to 30 to 60 MHz are possible using the present devices according to the FPGA vendors (the real range is dependent on application). In addition, the maximum system speed depends also on the results of the placement and routing tool in several device families. The largest FPGA devices available are about 10 kgates at present, when MPGAs provides more than 10 times higher usable logic capacity. Due to the considerably smaller gate capacity in FPGA devices, large designs have to be split into several devices decreasing the manageability of the design process and making the resulting system less reliable. For this reason, efficient partitioning algorithms and tools are needed. FPGAs provide usually quite a large number of I/O pins, several FPGA families much more than 100, which is quite useful, when implementing large systems and the design has to be split into several FPGAs. Especially with bit-parallel architectures, a large number of additional I/O pins might be needed due to the partitioning.

The FPGA implementation of the design is the matter of hours or days if no Printed Circuit Board (PCB) is needed to process. The corresponding processing time of MPGAs is normally several weeks (2 - 8 weeks) [12]. Also the hard competition between MPGA foundries and the fast turnaround Multi Project Chip (MPC) and Multi Product Wafer (MPW) type of services has lowered the NRE costs of silicon products adding pressure to FPGA market and making the selection between FPGA and MPGA more unclear in many cases. Basic technologies and FPGA versus MPGA cost comparisons are presented e.g. by Smith in [21].

In FPGA devices, the *programming technology* defines whether the device is programmable only once (antifuse), several times (EPROM, EEPROM) or "infinite" number of times (SRAM). The division of FPGAs according to the programming technology is presented with some examples in Table 1 [1]. Within the FPGAs, antifuse solutions, dielectric and amorphous silicon ones, provide remarkably better area effectiveness than the reprogrammable memory elements, but the total gate density cannot be stated exactly using purely the properties of programming elements, because the size of logic elements, the rest of routing area and the configuration logic also add the total area and they vary much in different architectures. The antifuses provide also the fastest interconnections due to the smallest internal resistance and capacitance, but the length of routings and the speed of actual logic element affects the overall performance. The volatile static RAM based FPGA needs an external memory device or computer cable for loading the configuration data unlike the other technologies, which can be used as standalone circuits.

	SRAM	Dielectric anti-fuse	Amorphous silicon anti-fuse	EPROM	EEPROM
Volatile	yes	no		no	
Programmable	infinite times	one time		several times	
Resistance (ohms)	500	250-500	25-100	1000	1000
Capacitance (pF)	50	2	1	10	20
Approximate area (μm^2)	50	1.5	1.0	10	20
Examples	Algotronix Plessey Xilinx	Actel TI *	Crosspoint QuickLogic	Altera Plus Logic Atmel	AMD NS Lattice

Table 1. Programming technology overview (* second source).

2.1 FPGA Architecture Overview

The FPGA architectures can be classified in two different ways, by the routing architecture and, by the granularity (size and flexibility) of logic cell. There exist roughly two different kinds of optimization goals. If programming elements are small then the use of the logic capacity in the architecture should be optimized, and vice versa, if the routing element is large, the architectural optimization has to emphasize more on the efficient use of routing elements. For the user there are several important parameters when selecting a suitable FPGA technology for the application: maximum operating speed, gate count, I/O count, price, implementation time, routing delay predictability, clocking possibilities and tool integration.

The *device architecture* limits the utilization of the nominal device capacity in several ways. It is very difficult to compare the logic capacities and maximal system speeds of different devices to each other, because those properties depend very much on the application and no good standardized comparison method exists. The easiest approach for the comparison would be that the FPGA vendors provide reference tables, (average performance and size required inside their devices) of the most commonly used structures. Currently, such information can be obtained from the conference papers or the application notes from the vendors. If there are not good reference cases, more detail studies of the architecture have to be done.

The present *routing architectures* [19] are based mainly on four different topologies: symmetric, asymmetric and global routing channels, and a sea-of-gates topology. In the symmetric and asymmetric channel architectures, logic modules are formed as a matrix of different sizes and connected via several routing channels between the logic modules. The routing results might be unpredictable due to the placement of logic cells. Those poor routing capabilities, especially in the case of symmetric routing channels, like in Xilinx devices, cause either unacceptable signal delays, or in the worst case even total failures in connecting the internal logic. However, with the reasonable placement of logic cells, such devices can provide high operating speed, which is limitted by the impedance caused by internal connections.

In the global routing channel solutions, multiple PLA and PAL like modules are connected to each other in the general interconnection area in the middle of the circuit. This approach provides normally quick routing to be completed in minutes, but the interconnection delays are neither very short nor very long. The almost constant interconnection delays make it easy to predict the speed performance of device, but at the same time it limits the maximum system frequency making them unsuitable for high speed ASICs. Of course, the single PAL type of devices provide very high speed logic, but those devices are quite small for their logic capacity for prototyping purposes in most cases. The sea-of-gates topology consists of an array of small logic modules covered by layers used for interconnections. That type of architectures do not support effectively medium and long distance routings, because there are no special routing channels for this purpose. For example, in the Algotronix device [2] the logic cells can be connected only to the nearest neighbour cells in the same horizontal and vertical line. The logic cells can also be used for routing purposes. Although it facilitates the routing, it will increase the routing delay and decrease the available logic capacity of the device.

The size of the logic cells varies a lot between different architectures from a pair of n- and p-channel transistors (Crosspoint) up to large combinatorial logic modules with two registers (ATV 2500 and 5000 from Atmel, and XC4000 from

Xilinx). A large logic module typically contains, in addition to a combinatorial part, a pre-defined sequential circuitry that can not be configured to implement any combinatorial function. The large logic module corresponds often to a stiff device architecture and it might affect dramatically the usability of device resources in some applications, but it also facilitates the estimation of capacity and performance in regular designs. Small logic modules can be configured to implement very simple combinatorial and sequential operations, or possibly they can be used for routing purposes making the circuit architecture more flexible.

One possible solution for logic formation is the look-up table based approach like in Xilinx devices. In such solutions, the content of the logic does not affect utilization of logic functions inside individual block, the only meaningful aspect is the number of needed inputs and outputs. The logic can also be implemented using basic gates such as in Concurrent, Crosspoint and Toshiba. Such solutions can be optimized for the narrow application to achieve high speed systems. In Actel and Quicklogic devices, all combinatorial logic is implemented using multiplexors. The most popular way of implementing logic functions is to use PAL like structures where a wired-AND plane feeds fixed OR-gates (Altera, Plus Logic and AMD). Those implementation techniques are quite often some way combined together to improve overall performance of the device architecture. Almost all logic blocks include at least one flip-flop or latch. It means that achieving high gate utilization in FPGA often requires designs which have about an equal number of combinatorial and sequential logic (the use of flipflop requires always 5-6 gates when the use of gates in logic modules is seldom so high). For example, the Xilinx devices have been found suitable for implementing regular DSP structures in [8] and also for strongly pipelined solutions in [15]. The number of low skew, high fanout drivers closely related to the device architecture can also be an important selection criterion at least when high operating speed requirements have to be satisfied.

One way to approximate the logic capacity of devices is to find out how the needed structures can be implemented in the tested architecture and logic cells. The quick comparison of the combinatorial logic can also be done by approximating the number of inputs of any Boolean function realized in one logic cell. The logic delays can be approximated using worst case delays from the databook, and if there are no guidelines to approximate the routing delay, it should be remembered that the routing delay might be bigger than logic delay. Some circuits have the fixed propagation delay, which facilitates the speed performance prediction, but the architecture also limits the maximum speed of the system. For example, Altera MAX5000 series devices have a fixed routing delay of 15 ns, while, for example, in Xilinx 3000 architecture the routing delays are normally some ns and for low fanout signals very seldom longer than 5 - 10 ns. When the net density increases, some high-fanout nets (over 5-6 loads) might have 40 - 60 ns routing delays due to the placement problems with the current design tools. In some circuits, additional control circuitry has been added to help logic minimization, for example, in MAX architecture the output of an OR gate is controlled by a XOR to allow Boolean minimization, but it increases the propagation delay, whether the control XOR function is needed or not.

2.2 Development Tools

Most FPGA development tools support the third party approach, which was the common factor of the initial strategies for the success of the current market leaders (Xilinx 56%, Altera 17% and Actel 14% in 1991 [1]) in the field of large FPGAs. Today, FPGA implementation systems support design entry from the popular design entry methodologies, like schematic capture, VHDL, Palasm, state machines and Boolean equations, and have translators from industry standards like EDIF 2.0.0. This facilitates integrating FPGA tools to the user's existing CAE environment and lowers the starting investments, because only a FPGA component library, a proper netlister and FPGA implementation tools are needed to complete the development system.

The mature current state of the art design system is based on either a schematic or a Register Transfer Level (RTL) VHDL synthesis tool. The abstraction level of design entry can be made higher and the corresponding time can be suppressed using application specific macro libraries like presented in [16] and [17]. Small changes in the coding style for synthesis tools affect dramatically the resulting implementation. Because the reasonable VHDL code for different synthesis tools differs, those libraries are tool dependent. With the macro libraries highly skilled professionals can also contribute their knowing even into the design of a "standard" product. For the design phase, Synopsys and Viewlogic synthesis tools, and the Viewlogic schematic editor have been used in our laboratory. The most matured tools (e.g. Viewlogic) provide the possibility to mix design environments so that the designer can either design with a schematic editor, or with a text editor and use block diagrams as an input of the schematic editor. Behind these boxes are the functional description of the design in a HDL-format. Higher level synthesis and SA/SD type of tools are not ready for the efficient design yet. Both synthesis tools with XACT design tool from Xilinx Inc. have been used for very fast prototyping, from synthesizable VHDL code to the measured prototype during even a single working day (1500 - 3000 equivalent gates). Some of the results are reported in [15]. A very general level product development environment is presented in Figure 2.

The partitioning of the design can be divided into three steps: the selection of suitable device, resource estimation, and the selection of partitioning criteria or algorithm. The needs of speed and actual logic capacity are quite difficult to approximate when using FPGA devices, because the utilization of the device is normally strongly dependent on the application. The emulators provide automatic partitioning tool for the design, but most current interface and implementation tools do not support automatic partitioning. When using manual design partitioning, the top-down approach by dividing the design into functional blocks might be the easiest way to avoid errors and to help the testing of the design [8]. When using the software approach for partitioning, also the bottom-up approach can be utilized [14]. At a moment, when synthesis tool is used for implementing the designs on FPGA, the partitioning should already be taken into account at the RTL level to assure the efficient utilization of device resources.

Some years ago, the only method in migration between technologies was the

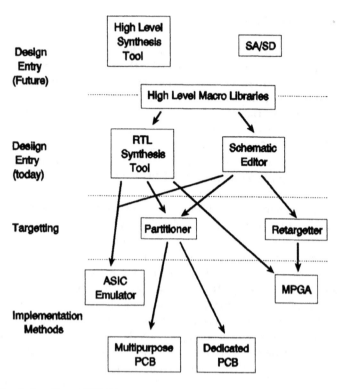

Fig. 2. The design flow of FPGAs.

manual redrawing of the design using the other target library, but even today universal links or retargetters do not exist. Therefore the technology alternatives and migration paths provided by the design tool should always be checked before making the selection of the prototyping technology. Generally, the migration from FPGAs to silicon technologies (mainly MPGAs) is provided directly by many vendors. For retargetting the schematic design from FPGA to MPGA, the *Library mapping* is normally used. Such a tool converts the individual cells in one library to one or several cells in another library with the cell function preserved in the conversion.

The synthesis tool uses *resynthesis*, where the characteristics of the netlist can be changed, normally resulting to a better result. At first, the library mapping of non-combinatorial cells to the target technology is done, and then the logic is decomposed into a technology independent format. In the resynthesis process of the decomposed logic, some logic minimizations can be done to optimize the resulting silicon area and performance. Today, this conversion needs some manual changes in VHDL source code. The Synopsys synthesis tool, for example, can be used to synthesize the technology independent RTL VHDL description down to the netlist, which can be automatically mapped to the selected FPGA component library. By now it supports Xilinx, Actel, Altera and Crosspoint devices. The quality of optimization for each technology can be checked after the technology mapping phase. To improve the quality, the synthesis tools provide a range of solutions according to the user requirements (trade-offs between speed, area and power) and they have usually a powerful ability to take into account constraints expressed in terms of critical inputs and outputs.

Although VHDL is an IEEE standard, the synthesizable subsets and especially the efficient coding styles for the syntheses tools differ so much that changing the design from one tool to another can cause several problems and possibly partial manual rewriting. The Viewlogic synthesis tool can not synthesize as large designs as Synopsys, which affects to the portability of designs. This can be avoided with very fine grained hierarchical design. Technology independency is much more harder to achieve. For example, Xilinx XC3000 series do not support asynchronous *set* in flipflops, and Viewlogic and Synopsys synthesis tools can not synthesize any code which requires this property, if there is not exactly such a component in the library. XC4000 series devices can implement asychronous *set* or *reset* if both are not needed in the same flipflop, instead Actel architecture, for instance, supports all those combinations. Therefore, the code has to be written for the final ASIC library. If an intermediate FPGA prototype is needed, the prototyping technology should be selected so that it supports the same properties as the target technology. Of course, such problems can be avoided with the careful design style taking into account both the possible prototyping and the final ASIC technology. For technology migrations, a fully synchronous design style should always be used to avoid mysterous problems caused by unpredictable routing delays inside FPGAs.

The routed netlist can be sent back to the simulator of the original design entry tool for delay simulation. If problems arise during the test phase, this property has been found vital in the design project. Unfortunately, at a moment, synthesis libraries for FPGAs do not support delay backannotated simulations. Because simulated and synthesized VHDL code might work differently, the synthesized netlist has to be verified against the high level specifications to guarantee the correct functionality. Such problems arise especially when using more than one synthesis tool during the project.

In addition to vendor's own implementation toolkits there are also at the market some device independent tools supporting the technology selection during FPGA evaluation facilitating the final ASIC migration. NeoCAD provides a FPGA foundry toolset [22], which is aimed at a device independent FPGA layout tool supporting by now only Actel and Xilinx devices. To achieve an efficient implementation on FPGAs some EDA vendors provide architecture specific synthesis tools for FPGAs, for example, AutoLogic synthesis tool (Mentor Graphics) [3] and Exemplar Logic [10]. The standard design libraries of FPGA vendors provide up to complexity level of large counters and bit-parallel adders. Currently, for example, Xilinx Inc. provides, in addition to the basic soft macros, also hard macros which are pre-partitioned to guarantee a predictable and higher performance. Today, high level generic macro packages are added to design systems to provide higher level components to speed up the actual design entry. For example, Xilinx Inc. provides X-blox, which is a module synthesis software using familiar design tools, such as Viewlogic schematic editor *Viewdraw* as a design platform [9]. It provides more than 30 parametrizable high level modules, like bus-wide boolean functions, adders, comparators and data registers.

3 FPGAs in an ASIC Design Project

Two possible approaches for the ASIC design project are based on Concurrent Engineering (CE) and fast prototyping concepts. In the concurrent engineering method, all environmental and market information is included into the design process at a very early stage to avoid the making of a prototype. For this purpose, system simulators provide a state of the art platform. Unfortunately, powerful mixed analog and digital simulators do not exist. Another major problem with the system tools is the lack of proper models, especially for non-electrical effects [11].

3.1 ASIC Design Project

The fast prototyping concept is usually used to demostrate the product as quickly as possible and that way minimizing the time needed to achieve the final specification of the product. If the specification cannot be certified properly beforehand, the configurable prototype using FPGA might guarantee a suitable platform to provide fast iterations. The configurable prototype is often seen as an additional side link, which delays the design of the "actual" ASIC. Using technology independent design tools or separate retargetters, for technology changes, for example, from the FPGA prototype to the actual MPGA implementation can be done automatically during few hours making the additional design delay unnoticeable. Anyway, the main target in FPGA prototype is to make sure the first time correct design on silicon. The prototyping concept is discussed in more detail in [6].

The design flow, which includes both approaches, is shown in Figure 3. The design of the product starts from the design idea, which has later been formed as a specification according to the initial design information. After the design step the product is implemented using suitable technology. What ever the selected technology is, the discussion concern cost, time, risk, resource and the knowledge of the project group. The design project is quite straightforward when there is enough design information (certified specifications) making the design flow near ideal CE model.

In every case, the time from the initial design idea to the design review should be minimized to improve cost effectiveness of the project team. If revisions are needed also recycle times have to be minimized. When speeding up the decision making process whether the product is rejected or accepted the useless work can be avoided. Of course, the rejected idea or project put into dustbin increase the amount of the design information for new projects and the project might be restarted later making the gained experience useful. The use of FPGAs in the design project lowers the threshold to implement the design ideas, thus providing more innovative products (usually better products), but at the same time they might add the number of rejected implementations causing additional work.

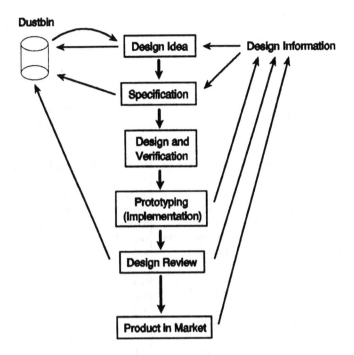

Fig. 3. The general product design flow.

The reuse of old information and components is the key factor for adding cost effectiveness of the design team and gaining profits. For this purpose, the automatic documentation of the implementation is valuable in the case of FPGAs, because short iteration times can lead easily to undocumented versions of design. The predictability of the design time is very important parameter when specifying an ASIC design project and to develop an accurate cost and time model for ASIC design. For this purpose, together with the backtracking on the previous projects, better automatic design tools are needed to minimize the human errors and that way to guarantee the predictable implementation time for the design.

The design flow chart does not discriminate the use of FPGAs as an end-product and as a prototyping device. For the product design project, as an concept, the implementation technology does not affect. FPGAs provide shorter idea to acceptance time lowering the threshold of processing novel ideas. They add more innovation to products by facilitating the experiments of novel application algorithms. Of course, they add at the same time useless prototypes and work making the production costs more expensive, but they add at the same time also the available design information and experience.

3.2 Utilization Approaches

For FPGA implementation, three basic approaches can be used: dedicated PCB, general multipurpose PCB and an ASIC emulator. The FPGA application can also be verified using multilevel system simulators. Those four approaches are compared in Table 2

	Dedicated PCB	General PCB	ASIC emulator	System simulator
Implementation time	slow	moderate	short	moderate
Design modifications	difficult	moderate to easy	easy	moderate to easy
Technology level	latest	new	mature	varies
Operating speed	quite fast	quite fast	moderate	very low
Complexity limitation	complexity management	complexity management	hardware capacity	simulation speed
Design (FPGA) management	manual	manual	automatic	manual
Reusability	difficult	easy	easy	easy
Total cost	moderate to high*	cheap	high to moderate*	high to moderate*
Human error propability	high	medium	low	moderate
Problems	PCB design time, manual design management	manual design management	old technology, only SRAM possible solution	model inaccuracy and availability, simulation slowness

Table 2. The comparison table of different FPGA utilization approaches (* depends on frequency of use).

The dedicated PCB solution is at its best when FPGAs have to be integrated closely to the other parts of the system. The largest problems of the approach are: the low tool integration level increasing the propability of human errors, requires an additional PCB design and the PCB solution is very difficult to utilize in future projects. A general multipurpose PCB [8] can be utilized in several projects. This lowers the error possibility in the FPGA implementation and speeds up the whole implementation phase. The better design management can be achieved by integrating the design and the implementation tools closely together. The bottle-neck in the design management is the partitioning of the design into several devices. Therefore the ASIC emulator based approach provides the most flexible and reliable approach for the implementation. The emulator provides usually lower maximum operating speed as the previous approaches due to the mature technology. The useability of the system simulator is limitted by the model inaccuracy and the slowness of the simulation with the long test vectors, and the mixing of analog device into simulations.

3.3 Suitable FPGA Solution

When using FPGA as an end-product, the designer is trying to find the smallest device to implement the design, but, in the case of ASIC prototyping, it is most important to get a working prototype in the minimum time with reasonable costs. In addition, the actual gate utilization of FPGA devices in the given application is very difficult to estimate beforehand, and the high gate utilization might cause problems in placement and routing [8]. Design changes during the prototyping phase might also cause a change of device, if device is too closely selected. In some prototyping cases, it might be reasonable to select more than one type of programmable device, for example, different types of devices for the high and the low speed parts of the design. This solution can be used to achieve the high operating speed goals with a reasonable number of devices, but it might also cause problems in implementation if the development software does not support both devices, and the overall design hierarchy has to be split to different tools.

The quality of the FPGA design tools and the way they can be integrated to the existing design environment are very important selection criteria. Poor FPGA development tools containing a lot of manual partitioning or routing may lead to unreasonable long development times. If the FPGA development tool supports several technologies, it gives more possibilities to the designer for reliable prototyping. For long term strategy selections, a flexible development platform is suitable selection, because the changes in the FPGA market can be very rapid and the current best FPGA device vendor might be a wrong selection one year later. However, the changing of prototyping technology and tools is not reasonable without very remarkable reasons, because it might take quite a long time to achieve the needed experience for reliable and optimized results, and new softwares and devices might cause additional unexpected problems. This aspect supports the selection of the reliable old vendor. Also, the new technology and tool add risks in the project scheduling, but it might still be the best solution. Anyway, all softwares as well as devices should be available at the time of technology selection, because "coming soon" might mean several years or never, and it has to be remembered that software tools are normally delayed as compared to corresponding FPGA devices. FPGA development tools might need a very powerful design platform and flexible memory operations, therefore UNIX-based workstation should be supported, especially when devices with large logic capacity are used.

One possible selection for a prototyping tool is a hardware emulator [7], which can be integrated directly into the actual ASIC development environment. The emulators are quite competitive tools when the designer does not have proper knowledge about the prototyping technology and the target ASIC technology is considered beforehand. The emulators can also be used effectively with HDL based synthesis tools providing a rapid prototyping of the design. Unfortunately, the emulator is quite an expensive solution, but it might speed up and facilitate the prototyping so much that it is worth of its cost. It is not suitable for rapid prototyping when small power consumption and size are needed.

4 Future Trends

Very hard competition with MPGAs and on the FPGA market force the prices of usable gates to sink all the time. For example, the prices of 1000 gate FPGAs have decreased by a factor of two or three since 1989. This development is supposed to continue with light saturation due to the technological limits. This price reduction is also expected to happen for the new generation of complex FPGAs in few years. Current FPGA price is 3 to 4 times the price of similar capacity MPGA in low quantities. Although the FPGA market is changing very quickly all the time, and there is an increasing number of vendors and not enough markets for all of them, the current market leaders: Xilinx, Altera and Actel are forecasted to stay on the strong position [13]. Due to the enormous number of different interface formats, the third party interfaces have to become more flexible using standards (e.g. EDIF and VHDL) to sink the software development costs and time.

The market price of the FPGA is a function of the die complexity and the number of pins needed (package cost). The gate price is reduced proportional to the reduction of cell area in CMOS technology. Although the size of design which is reasonable to implement with FPGA becomes larger, it means that no dramatical changes in the break-even point (volume) between FPGA and MPGA are expected. Because the packaging cost of device is normally several times higher than die price, the role of I/O pins will become more important. For ASICs, up to 450-500 package pins are needed in 1995 [5]. According to Rent's rule the number of needed I/Os depends on the usable gates [23]. Thus as the average size of ASIC will increase also the average number of I/O pins will increase. Because usually several FPGAs are needed, the required pin count might be much higher than the corresponding MPGA. It means that more expensive packages (Quad Flat Packs and Pin Grid Arrays) are needed for FPGAs than MPGAs. Also the window based package (EPROM) is quite a remarkable cost factor.

In addition to that, device properties (current drive capability and input capacitance of transistors), on-chip interconnections, chip-to-chip interconnections and cooling capacity are limitting the performance of the device. As the major limitation in the integration ratio and the performance in bipolar devices is the power dissipation density, the integration ratio on a CMOS chip is limitted by the die yield. The availability and the price of most advanced FPGAs are defined by the yield. According to the forecast development of CMOS technology (die edge will increase about by the factor of two, limitted by the yield, and the dimensions are decreasing about by the factor of four in every decade) the maximum gate count of FPGA devices will be about 40 kgates and the system speed around 100 MHz by 1995. BiCMOS technology combines the low static power consumption and logic flexibility of CMOS with the high speed, low logic swing, and input signal sensitivity and high output drive of bipolar technology. The processing cost is higher due to the greater number of processing steps. BiCMOS devices are supposed to be about 50% to 100% faster than CMOS ones [1]. This development may be enhanced by new innovations in programming technology and their integration to standard ASIC technology.

Most ASIC designs are in the speed area of 20 to 100 MHz and the maximum ratings in the telecommunication are in the range of GHz [4]. It means that the speed requirements are mainly beyond the capabilities of FPGAs. But for a rather low speed DSP from instrumentation to digital audio applications, programmable devices should retain their competitiveness. Therefore the development of FPGA architectures is application driven and going towards application dedicated architectures and hard macros to achieve reasonable performance. Problems with logic capacity can be overtaken using emulator type of approach, but the high speed goals cannot be achieved by connecting several devices.

Generally, the FPGA market will be customer support driven instead of technology. In future, the success of FPGA companies will be determined by the customer acceptance. Critical factors for this goal are robust and functional FPGA vendor specific tools and working interfaces to main stream design front-end tools such as synthesis and schematic capture. Without this, FPGA vendors are doomed to fail in future as they have been in the past. Other critical aspect for FPGA vendors is how to handle the transition from FPGA to MPGA. Naturally, the interest of the FPGA vendor is to sell FPGA circuits in large quantities to the customers. However, the customers are interested to use the most economical solution, which will satisfy their technical and time-to-market requirements.

5 Conclusions

The rapid increase in the capacity and the performance of FPGAs has been made them a suitable solution for ASIC prototypes or even end-products when flexibility and short development time are emphasized. As the abstraction level of the design entry has also increased and development tools are becoming more flexible, the actual implementation technology does not affect the design entry phase, and the knowledge of the differences between various FPGA families are not so significant any more. As programmable technologies, FPGAs give new possibilities for the design teams to utilize their knowledge and innovation.

In addition to the maximum operating speed and logic capacity, three other factors have to be taken into account when a selecting suitable FPGA family for the application: device architecture, programming technology, and the development tools. To provide the proper comparison of FPGA solutions, the reference table of the most commonly used structures covering the essential implementation properties: operating speed and capacity, would be the best solution.

References

1. *The Programmable Logic IC Market - Application, Technology and Market Trends.* Electronic Trend Publications, Saratoga, CA, USA, 1992.
2. C. Carruthers and T. Kean. Bipolar cal chip doubles speed of fpgas. In W. Moore and W. Luk, editors, *FPGAs*, pages 46—53. Abindon EE&CS Books, Abingdon, England, 1991.
3. Mentor Graphics Corp. *FPGA Optimization with AutoLogic FPGA*. San Jose, CA, 1991. Vendor Datasheet.

4. S. Fox and P. Gordon. *The ASIC and Programmable IC Strategy Report*. Electronic Trend Publications, Saratoga, CA, USA, 1991.

5. DM Data Inc. *Pricing of ASIC Parts - 1991*. Scottsdale, Arizona, USA, 1991.

6. J. Isoaho. *DSP ASIC Development Based on Prototyping and Bit-Modelling*. Licentiate Thesis, Tampere University of Technology, Tampere, Finland, May 1992.

7. J. Isoaho, J. Pasanen, A. Nummela, and H. Tenhunen. DSP ASIC evaluation with fast prototyping. In *Proc. EURO-ASIC92*, pages 102—106, Paris, France, June 1992.

8. J. Isoaho, J. Pasanen, O. Vainio, and H. Tenhunen. DSP system integration and prototyping with FPGAs. *Journal of VLSI Signal Processing*. To be published 1993.

9. S. H. Kelem and J. P. Seidel. Context-based ASIC synthesis. In *Proc. EURO-ASIC92*, pages 226—231, Paris, France, June 1992.

10. L. Maliniak. Synthesis tools move into the mainstream. *Electronic Design*, pages 51—64, Aug. 1991.

11. L. Maliniak. System simulation still holds promise. *Electronic Design*, pages 53—61, Feb. 1992.

12. W. J. McClean. *ASIC OUTLOOK 1992 An Application Specific IC Report and Directory*. Integrated Circuit Engineering Corporation, Scottsdale, Arizona, USA, 1991.

13. W. J. McClean. *STATUS 1992 A Report On The Integrated Circuit Industry*. Integrated Circuit Engineering Corporation, Scottsdale, Arizona, USA, 1991.

14. W. O. McDermith. A bottom-up approach to FPGA partitioning. In *Proc. IEEE 1992 Custom Integrated Circuits Conference*, pages 5.4.1.—5.4.4., Boston, USA, May 1992.

15. J. Nousiainen, J. Isoaho, and O. Vainio. Fast Implementation of Stack Filters with VHDL-Based Synthesis and FPGAs. In *Proc. IEEE Winter Workshop on Nonlinear Digital Signal Processing*, pages 5.2-4.1—5.2-4.6, Tampere, Finland, Jan. 1993

16. J. Nousiainen, A. Nummela, J. Nurmi, and H. Tenhunen. Strategies for Development and Modelling of VHDL Based Macrocell Library. In *Proc. The European Conference on Design Automation with The European Event in ASIC Design*, pages 478—482, Paris, France, Feb. 1993.

17. A. Nummela, J. Nurmi, J. Isoaho, and H. Tenhunen. Strategies for implementation independent DSP system development using HDL based design automation. In *Proc.ASIC'92*, Rochester, USA, Sept. 1992.

18. B. Prince. *Semiconductor Memories - A Handbook of Design, Manufacture, and Application (Second Edition)*. West Sussex, England, 1991.

19. J. Rose, K. El-Ayat, C. McCarthy, and S. Trimberger. Field-programmable gate arrays. In *Design Automation Conference Tutorial*, June 1991.

20. G. Saucier, E. Read, and J. Trilhe, editors. *Fast-prototyping of VLSI*. Elsevier Science Publishers B.V., Amsterdam, Netherlands, 1987.

21. M.J.S. Smith. ASIC Technologies. In *Proc. Fifth Annual IEEE International ASIC Conference and Exhibit*, Rochester, USA, Sept. 1992.

22. B. Tuck. Designers search for the secret to ease asic migration. *Computer Design*, pages 78—117, Dec. 1991.

23. R. R. Tummala and E. J. Rymaszewski. *Microelectronics Packaging Handbook*. Van Nostrand Reinhold, New York, USA, 1989.

Some Considerations on Field Programmable Gate Arrays and Their Impact on System Design

Alberto Sangiovanni-Vincentelli
Department of Electrical Engineering and Computer Science
University of California at Berkeley

Abstract. User programmable devices are becoming more and more important for system design because of their flexibility that allows much shorter design time and hence better time-to-market. The use of programmable devices spans rapid prototyping and dynamically reconfigurable systems. The market is projected to raise rapidly to levels that make this type of devices very appealing for merchant IC manufacturers. Some of the issues related to the future of these devices as well as to the future of system design methodologies are presented.

1 Introduction

Historically, the electronic industry has been partitioned into two main segments:

- the Electronic System industry that manufactures complex apparati sold to a final customers to perform a specific function. Typical examples are computer companies such as Apple, DEC, telecommunication companies such as Fujitsu, Siemens, and consumer electronics companies such as Thomson, Sony, Matsushita.
- The semiconductor industry which manufactures the components needed to implement complex systems. Typical examples are AMD, Intel, SGS-Thomson, National.

At times, the two are divisions of the same parent company for example in vertically integrated companies such as IBM and most of the Japanese electronic industries. Even in this case, the strategic considerations that guide these groups are different. In fact, a semiconductor manufacturer has been traditionally motivated by and interested in high-volume products that make its capital investments more productive, while a system company is mostly interested in serving a diversified market where customization is desirable and where time-to-market is the essence.

In the past, the semiconductor industry provided electronic components that were intended for large markets cutting across application areas. Typical examples were memories and standard logic products. Competition in this arena was on manufacturing excellence and cost. Yielding to the pressure of more sophisticated system designers, semiconductor companies started developing Application Specific Standard Parts (ASSPs), components which are more specific to a particular application and have smaller sale volume. The smaller sale volume was compensated by higher prices that these components could fetch in the market because of the value added in the design of the components.

Nevertheless, the most successful ASSPs were still produced in fairly high volume and served more than a single user. In the 1980s, the advent of more flexible manufacturing and of effective CAD tools such as simulators, and place and route programs, made it economical for system companies to design their own semiconductors to obtain better and better performance. A new semi-conductor business model was born: the semi-custom or User Specific Integrated Circuits (USICs) model. The success of a company in this sector was basically dependent upon service and turn-around time. However, the price that could be charged for these components was not as attractive for a semiconductor company as the one of ASSPs since most of the value added was in the hands of the system companies. The most interesting revenue source was the NonRecurrent Engineering costs (NREs) that were paid for the services needed to complete a design successfully.

Designing their own semi-conductors did make the system companies more competitive in terms of product performance but this higher performance was paid by fairly long turn-around time (design errors required to design and manufacture a new chip) and by a consequent longer time-to-market. Today, increasing pressure to bring to market new products and the more rapid obsolescence of existing products makes designing semi-custom circuits more and more problematic even if fairly new design aids such as logic synthesis are improving significantly designers' productivity.

The need to enter the market with innovative products as soon as possible also implies the need of adjusting quickly products that do not meet customer expectations. This in turns implies design methodologies that are flexible enough to accomodate engineering changes very rapidly. The introduction of Field Programmable parts such as PLDs and PALs allowed very short turn-around time and great flexibility, but the size of designs that could be accomodated was still too small for most of the complex electronic systems [1]. However, the introduction of Field Programmable Gate Arrays allowed the use of design styles similar to the ones followed in traditional ASICs.

The trend is clearly towards the more and more extensive use of programmable devices. I believe that most of the future system designs will be centered around a *software programmable device* such as a micro-controller, a micro-processor or a digital-signal processor and will contain communication and glue logic implemented by Field Programmable Gate Arrays.

Only in few cases the use of semi-custom circuits will be justified from performance point of view. In fact, the performance of micros is already so high that it is difficult to imagine a situation where they not be fast enough to perform most of the tasks needed in the system applications. In addition, the performance of Field Programmable Gate Arrays is improving constantly; consequently the gap with semi-custom devices is decreasing even though a gate-array solution is still more than 5 times faster and up to 3 times smaller than the equivalent FPGA

---

[1] PALs and PLDs implement two-level logic that may not be effective to represent complex Boolean expression since it does not exploit common subexpressions.

implementation [1].

However, if the number of units to be manufactured is large, the price per component of FPGAs is high enough to make their use in final system production expensive. On the other hand, NREs are basically non existent for FPGAs, and in small productions, FPGAs are also appealing from a cost point of view. There are reasons to believe that soon the prices of FPGAs would be much lower, since manufacturing costs for FPGAs should be comparable to the ones of memories.

All market projections and indicators are signaling the extraordinary growth potential of FPGAs. For example, the number of designs that make use of these components is now larger than the number of designs using USICs and it will eventually be much larger (see Figure 1). In addition, the growth of the mar-

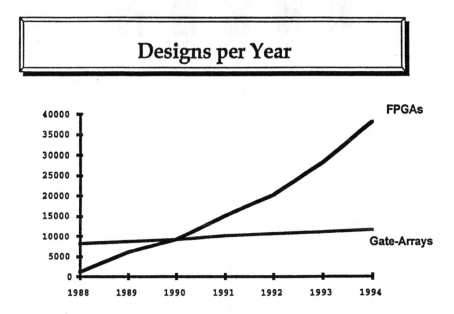

Fig. 1. FPGA Design Starts vs. Gate-Arrays

ket in terms of dollars is staggering (see Figure 2). FPGAs are very appealing products: on one hand, they allow flexibility, fast time-to-market, customizability, just-in-time production, on the other they are "standard" parts and can be manufactured in high volumes.

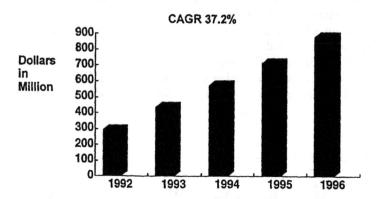

Fig. 2. Market Projections

This paper addresses some considerations on the use and the industrial future of FPGAs. In particular, in Section 2, the application market for FPGAs is briefly investigated. In Section 3, considerations on the futu e of the industry are introduced. In Section 4, conclusions are offered.

2 Field Programmable Gate Array Market Segmentation and Evolution

The FPGA market is evolving along three main directions:

- Rapid prototyping, where FPGAs are used to obtain a "bread-board" of the system being designed. In system design the use of simulation is increasing, but simulation speed is often not enough to provide the debugging capabilities needed to verify extensively the performance and the functionality of the design. Validation techniques such as formal verification and emulation are now being introduced to complement but not replace the use of "breadboarding". FPGAs made it possible to build general purpose emulation systems. In fact, general purpose emulation machines such as Quickturn and PIE, are

built with a large number of re-programmable FPGAs. For rapid prototyping, the most desirable feature is flexibility (and hence re-programmability) so that the bread-board could be easily modified to correct design errors.
- System implementation, where FPGAs are parts of the final design. The most desirable features here is manufacturing turn-around time and price/performance In general, higher price/performance ratios for FPGAs versus semi-custom are compensated by time-to-market advantages.
- Dynamically reconfigurable subsystems, where FPGAs are used to implement different functions that are dynamically loaded into the array from an external agent such as a micro-controller or a micro-processor. This application is perhaps the most innovative for FPGAs and it is still in its infancy. Network protocols, universal hardware co-processors [4], workstations [3] are some of the most exciting applications. In this case, the desirable features are re-programmability and speed in loading the array from external sources [5].

Originally FPGAs were used for rapid prototyping mainly. However, system designers realized that fast turn-around was essential also for the final version of their designs and, in fact, the largest number of parts today is sold to implement final versions of the system. Dynamically reconfigurable systems are really fascinating possibilities and their market is still largely unexplored.

Each market has its own ideal device. Today, SRAM-based solutions are dominating the re-programmable market, while anti-fuse-based solutions are present in the high-performance market. For dynamically reconfigurable subsystems, memory-based solutions are favored [4, 5, 3]. However, more research is needed to couple FPGA architectures with algorithms.

It is doubtful whether a single solution will exist for the three markets. I believe that three different product lines will dominate the respective markets.

3 Trends: Business Strategies and Design Tools

3.1 Business Strategies

FPGAs are standard parts: consequently, they should have the same price dynamic as memories.

Today the market is dominated by relatively small fab-less companies such as Xilinx, Actel and Altera. Their products are manufactured by foundries offered by large semi-conductor companies. The price of FPGAs must take into account the value added by the design of the FPGA companies and the manufacturing costs of the foundries.

Given the projected growth of the market (see Figure 2), it is conceivable that the major semiconductor companies will enter the FPGA market in a major way. ATT is already manufacturing and selling SRAM-based FPGAs and Toshiba has developed an FPGA in collaboration with Pilkington Glass. Matsushita and Texas Instruments have manufacturing and distribution rights for Actel parts. The market share of the FPGA producers is shown in Figure 3.

The following questions arise:

FPGA Market Share

Rank		Supplier	Dollars		% Share	
`91	`92		`91	`92	`91	`92
1	1	Xilinx	130	161	72	65
2	2	Actel	37	45	20.5	18.2
3	3	TI	6.4	18.8	3.5	7.6
4	4	ATT	6.2	17.5	3.4	7.1
5	5	QuickLogic	0.3	2	0.2	0.6
-	6	Crosspoint	0.0	2	0.0	0.8
-	7	Concurrent	0.0	0.5	0.0	0.2
6	8	Plessey	0.2	0.1	0.1	0.0

Fig. 3. Market Share for FPGA Producers

— *Will the major companies continue to license the products of the fab-less companies or will they develop their own products?*
— *Will the FPGA companies survive in the long run?*

The protection offered small manufacturers by their patent portfolio is key for their survival. Yet large semiconductor companies own an extensive patent portfolio covering all aspects of IC manufacturing and design that could be used to counter any law suit of the FPGA companies. In addition, the shear revenue size of large IC manufacturers allows to weather out expensive law suits relatively unscathed. Thus I believe that eventually the market will open up. In my opinion, the survival of the FPGA companies is pretty much related to their flexibility, innovation capabilities and installed basis. New architectures and products developed by FPGA companies are coming to market at a good pace making it relatively difficult for the major semiconductor houses to keep up. However, there will be a time where innovation will not provide enough advantage for the small companies. Stacked profit, i.e., the fact that the fab-less companies have to pass to the foundries part of their profit, will become relevant.

Two avenues are then open for the FPGA companies: acquisition by a major semi-conductor company or mergers among the main competitors. Today the FPGA market is about $400M per year, of which Xilinx is about $200M, with estimated net income of about $30M, but estimated to grow to approximately $1B by 1997, (see Figure 2 and 3) The combined revenues of the small FPGA companies will probably be enough to allow the set-up of independent manufacturing capabilities thus freeing FPGA companies from the dependence from foundries.

Installed basis is strategically very important. Today, the use of FPGAs is not straightforward. System designers are reluctant to use different development systems and architectures. However, new design methods and tools may overcome the fear of innovation of system designers, traditionally a conservative group very resistant to changes, especially when compared to IC designers.

3.2 Design Tools

Design systems are key to market penetration and extended use of FPGAs. In fact, it is of little use to have a part that can be programmed in few hours but that takes weeks to months to design. Today, design systems excluding design entry, are sold mostly by FPGA companies. The design systems are proprietary and do not allow to move freely from one FPGA manufacturer to another. However, standard formats appeared that allow to use a unified representation of a design that could be mapped into a variety of devices.

Proprietary systems focus mainly on place and route tools since FPGA companies control the actual physical layout of the device [1]. Timing analysis tools are also provided by the vendors. However, while place and route tools are essential to obtain fast design and turn-around time, they are not sufficient to make the design time of FPGAs short. In particular, given the complexity of the logic blocks of the most commonly used FPGA architectures, logic design is non trivial. Several design groups in large companies with good design tools used logic synthesis to map designs into FPGAs. However, this strategy was not very effective because the particular architectures of FPGAs were not exploited to their full power, since the logic functions implementable by the blocks were expressed in terms of standard ASIC libraries [2]. Some of the FPGA vendors developed their own synthesis tools or OEMed synthesis tools provided by a third party.

The following questions arise:

- *since design systems are a good revenue source will the FPGA companies continue develop and support design software? and if yes, which ones?*
- *When will the dominant CAD companies enter the FPGA tool market in a major way?*

The main issue to be faced by CAD vendors is whether the price they could charge for their FPGA tools is worth their development effort. The PAL and PLD design tool markets are characterized by very low prices since system designers using these devices do not have a sizable budget for tool acquisition. However,

FPGAs are really a different market altogether: they are somewhat in between the traditional Programmable Logic market and the gate array market. In fact, several large companies such as HP are making the choice of FPGAs to be used in system design a corporate choice. Once the FPGA design methodology permeates these large companies, the price that could be charged for design tools is definitely higher. In addition, these large companies require the development of what is called *Technology Transparent Design, TTD* since it is in their best interest to develop a design in a technology independent fashion and then use the best FPGA device to implement it in terms of costs and performance. In this paradigm, it should also be possible to migrate a design from an FPGA to a gate-array or to a standard cell chip. In TTD, the use of synthesis tools is absolutely essential. Exemplar logic pioneered the development of logic synthesis tools for FPGAs, Synopsys recently announced a product for the direct synthesis on FPGAs. I expect the area of synthesis for FPGAs to grow by leaps and bounds. However, since the design community for FPGAs is much more diverse than the ASIC design community and cost is a real issue, I expect a new breed of FPGA synthesis tools that are almost transparent to the user, even though some quality will definitely be lost, and low cost.

As far as layout is concerned, according to the TTD paradigm, there is also a trend to move away from proprietary systems so that a designer could choose with relative ease one implementation over another in the same design system. Neocad and Cadence announced products that allow the physical mapping of a design into the most common FPGAs. Given the recent proliferation of architectures, it may be too difficult and time consuming to develop a real TTD tool [2]. Hence there is a definite probability that successful tools will target only the most popular architectures. This may facilitate the expansion of FPGAs into new design groups but may provide a high barrier to entry for new architectures.

4 Conclusions

FPGAs are the most exciting component in the ASIC family today. I believe that FPGAs will penetrate all aspects of system design in a major way. I expect new architectures to be introduced shortly with one or more of the following features:

1. coarse and fine grain logic blocks,
2. routing channels will disappear (at least in the anti-fuse family) similar to the sea-of-gate concept,
3. FPGA blocks will be included onto a micro-controller chip to provide special user-specific functions that can be even modified dynamically.

There will be great value in offering memory-based and anti-fuse-based FPGAs that are consistent in terms of design tools and design entry systems. Major semi-conductor companies will be aggressive in the FPGA market. The present

---

[2] Even though the Exemplar Logic tool set targets almost every device being marketed.

FPGA vendors will have to develop a strategy that will make their survival possible even in the presence of strong competition.

CAD vendors will enter the FPGA arena in a major way. Consequently, there will be a re-direction of the companies that develop internal design software to leverage maximally the expertise and the sale channels of the CAD vendors.

FPGAs will be used for new produts in many different ways that will make system design simpler and more flexible. FPGAs will also be used in design courses in every learning institution. In short, FPGAs will affect every corner of the electronic industry.

5 Acknowledgements

This paper is the distillation of many conversations with a number of FPGA producer and experts, John East, Abbas ElGamal, Jonathan Rose, Steve Trimberger, Bob Ko, Richard Newton. This work is partially supported by DARPA under contract numbers J-FBI-90-073.

References

1. J. Rose, A. E. Gamal, and A. Sangiovanni-Vincentelli. A Classification and Survey of Field-Programmable Gate Array Architectures. In *Proceedings of the IEEE*, Sep. 1993, to appear.
2. A. Sangiovanni-Vincentelli, A. ElGamal, and J. Rose,. A Classification and Survey of Field-Programmable Gate Array Architectures. In *Proceedings of the IEEE*, Sep. 1993, to appear.
3. Beat Urs Heeb. Debora: A System for the Development of Field Programmable Hardware and its Application to a Reconfigurable Computer. In *Doctoral Dissertation for the Swiss Federal Institut of Technology Zurich*, 1993.
4. P. Bertin, D. Roncin, and J. Vuillemin. Programmable Active Memories: a Performance Assessment In *Proceedings of the 1993 Symposium on Research on Integrated Systems*, pp. 88-102, 1993.
5. D. Poutain, Algotronix: The First Custom Computer In *BYTE*, Sep. 1991.

SRAM-Based FPGAs Ease System Verification

Bradly K. Fawcett
Xilinx Inc.
2100 Logic Drive
San Jose, CA 95124

Abstract. Today's system design teams are faced with the conflicting challenges of greater system complexity, shorter development cycles, and stringent quality and reliability requirements. Many designers are meeting these challenges through the use of Field Programmable Gate Arrays (FPGAs). Fully-tested, static-memory-based FPGAs can be configured and reconfigured while resident in the target system, providing an efficient prototyping vehicle for verifying custom ASIC designs, as well as a fully-tested device for production use. Their reconfigurable nature allows for the cost-effective inclusion of self-diagnostic test logic in systems.

SRAM-Based FPGAs

Since their introduction in 1985, Field Programmable Gate Arrays (FPGAs) have been used successfully in thousands of applications, and are the fastest growing segment in the digital logic market today. Combining the density and flexibility of a gate array architecture with the convenience and time-to-market benefits of a user-programmable device, FPGAs provide designers with a cost-effective ASIC device for both prototyping and production use. FPGA technology is having a profound effect not only on the way that users design and build electronic systems, but also in the way those systems are tested and verified.

The growing complexity of electronic components and systems has increased the need for fast, accurate, and cost-effective device and system verification techniques. These complexities extend both to the system's logic design, with increasingly fast and dense circuits being incorporated into large ASIC devices, and the system's mechanical design, with surface-mount technologies, high-pin-count packaging, and multilayer PC boards. These new technologies often outstrip the capabilities of traditional verification tools, such as chip-level simulation and "bed-of-nails" board testing.

These complex testing problems can be eased by the use of static-memory-based Field Programmable Gate Arrays. In these devices, the state of internal static memory cells determines the logic functions and interconnections resident within the FPGA device. By downloading different configuration programs into the memory cells, the device's function can be changed at will while the FPGA is resident in the target system. This extreme flexibility facilitates system verification in several ways. SRAM-based FPGAs can provide a fully-tested, high-density ASIC device for production use in a system. FPGAs also can provide an effective prototyping tool to verify the logic of a design, especially circuits targeted for implementation as a custom ASIC device. SRAM-based FPGAs also provide an easy, cost-effective mechanism for including self-diagnostic logic within a system design.

SRAM-based FPGAs were first introduced by Xilinx Inc. in 1985. The Xilinx Logic Cell Array (LCA) architecture features three types of user-configurable elements: an interior array of logic blocks, a perimeter of I/O blocks, and programmable interconnect resources. Configuration programs can be loaded automatically at power-up or upon command at any time. The programming data itself resides external to the FPGA, typically in an EPROM or ROM, or on a floppy or hard disk. Several available loading modes accommodate various system requirements. Devices can be programmed and re-programmed an unlimited number of times. (Other vendors of SRAM-based FPGAs include AT&T Microelectronics, Altera, and Atmel.)

The benefits of a static-memory-based FPGA include high-density, high performance, testability, and the flexibility inherent to a device that can be programmed while resident in a system. "Soft" hardware can be implemented; that is, changes can be made to a system's logic functions simply by reconfiguring an FPGA, without physically altering the printed circuit board in any manner. Often, systems will include multiple configuration programs for their SRAM-based FPGAs, allowing varying operations - including testing and diagnostic operations - to be efficiently performed with a minimal amount of hardware.

A Fully-Tested ASIC

System-level verification cannot occur unless the operation of the individual components of the system also is verified. Device testing issues add significantly to the complexity, time, and cost of custom and semi-custom ASIC design. Verification of a custom ASIC requires, as a minimum, the development of test programs and test vectors. However, user-created vectors seldom provide more than 80% fault coverage, and automatic test vector generation (ATVG) software typically provides just 90% fault coverage. Often, test circuitry must be included in the design, such as scan-path test logic, to move fault coverage above 90%.

Alternatively, SRAM-based FPGAs can provide the same density and architectural flexibility as custom ASICs, while avoiding these expensive verification problems. SRAM-based FPGAs can be fully tested by the manufacturer. For example, Xilinx FPGAs are designed so that each functional node can be configured and routed to outside pads. Short configuration loading times (< 100 milliseconds) allow multiple configuration patterns to be loaded without removing the device from the tester. All configuration memory bits, all logic elements, and every metal line is exercised and fully tested. 100% fault coverage is guaranteed.

(In contrast, EPROM- and antifuse-based FPGAs cannot reach this level of fault coverage. Due to the long erase times of EPROM cells and the one-time-programmable nature of antifuses, these programming elements cannot be exhaustively tested by the manufacturer. Some post-programming verification is required, and less than 100% programming yields can be expected.)

Thus, the use of SRAM-based FPGAs in place of other ASICs reduces the user's design time and effort, since the designer does not have to be concerned about testability requirements during the design cycle. The SRAM-based FPGA concept not only removes the burden of test program and test vector generation from the user, but also removes the question of fault coverage and eliminates the need for fault grading. The FPGA is a standard part that guarantees any valid design will work.

FPGAs and Prototype Verification

While many users employ FPGAs in their production products, others use FPGAs as a prototyping tool during semi-custom or custom ASIC design. Either way, the FPGA provides an efficient, cost-effective vehicle for testing and debugging a system's logic functions.

SRAM-based FPGAs provide a flexible means of "breadboarding" logic designs. Configuration programs can be downloaded to an FPGA resident in the target system, facilitating the testing and debugging of the FPGA's logic while operating in the actual application, and the verification of total system operation. This type of in-circuit testing has several advantages over software simulation. No simulation vectors are required, and the system can be tested at operating speeds. The interface between the FPGA and the other system components can be fully verified. The scope of the testing can be extended to all possible system operations, not just those modeled by the limited set of test vectors typically used for simulations.

Design iterations can be quickly generated and tested using SRAM-based FPGAs, encouraging experimentation. Likewise, temporary modifications to the logic for debugging purposes, such as routing an internal node to an otherwise unused I/O pad, are easily implemented. Usually, implementing a small change and downloading a new configuration program for in-circuit verification requires only a few minutes effort. Prototyping expenses are minimized, since the FPGAs are infinitely reusable, and logic can be modified just by downloading new configuration programs. There is no lengthy wait for a custom device to be manufactured, and no waste of components as with one-time-programmable solutions; there is not even the inconvenience of removing the FPGA from the target system for erasure and re-programming, as with EPROM-based logic.

While in-circuit testing holds many advantages over software simulation, simulation or static timing analysis is still required to verify system timing under worst-case conditions. However, if the FPGA's functionality has been verified in-circuit through actual system operation, then the exhaustive simulations typically associated with ASIC development are unnecessary; only the critical timing paths need to be analyzed.

Once the correct operation of the FPGA-based prototype is verified, the designer must choose the appropriate technology for implementing the logic in production systems. If the speed and density requirements of the design can be satisfied with FPGA devices, then the choice between FPGAs and custom ASICs is largely an economic one. The potential lower unit price of the custom device has to be weighed against the cost and risk of converting the design from an FPGA to a custom technology. "Opportunity costs" also must be evaluated; that is, the trade-off between using engineering time to convert the design as opposed to starting development of a new product.

For high-volume applications, Xilinx offers another alternative - the HardWire family, a mask-programmed version of the programmable FPGA architecture. A HardWire device is architecturally identical to its SRAM-based equivalent, providing pin and performance compatibility. The conversion from the FPGA is transparent. An on-chip scan test path in the HardWire device allows for automatic test program generation with 100% fault coverage. Unlike conversions to other ASICs, virtually no engineering resources are required for a HardWire conversion; neither logic redesign, test vector generation, nor PCB redesign are required.

Even for ASIC implementations that exceed the density and performance of today's FPGA devices, prototyping with SRAM-based FPGAs still offers significant benefits. In fact, several vendors, including Quickturn Systems (Mountain View, CA), PiE Electronics (Sunnyvale, CA), and Integrated Circuit Applications Ltd. (Berkshire, UK), offer ASIC emulation systems based on Xilinx FPGAs. These systems are composed of multiple SRAM-based FPGAs that can be configured to implement any given logic functions; designs comprising tens of thousands of gates can be emulated. With this scheme, system functionality and performance may be tested and revised before committing the ASIC to silicon, greatly reducing design risks. There is no silicon "waiting period" in the critical path to verify overall system operation. Internal nodes may be accessible for debugging that would not be available in the custom ASIC implementation. Software development can continue while the silicon is being manufactured. The FPGA-based emulator could even be used for early demonstration of the final product to potential customers.

FPGAs and System Self-Diagnostics

Built-in, self-diagnostic test logic assists the designer during initial system verification, increases the confidence level of the system's users, and aids in field maintenance operations. However, using 'traditional' logic technologies, the additional logic needed to implement diagnostic operations can significantly increase the system's size, complexity, and cost.

Reprogrammable SRAM-based FPGAs offer the designer an economical means to implement test logic without adding to the hardware content of the system. The same FPGAs that implement the system's logic can be reconfigured while in the system to hold diagnostic logic to test that system. When the system is powered-up or placed in a test mode, its FPGAs are configured with logic functions dedicated to testing other circuitry in the system. Once the testing is successfully completed, other configuration programs are loaded into the FPGAs to implement the actual logic of the particular end application intended for that system. The diagnostic test logic is essentially "free", except for the additional memory space required for the diagnostic configuration programs.

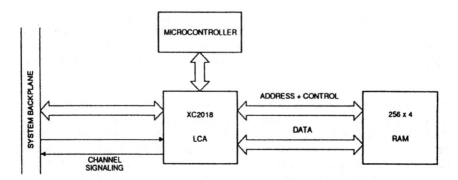

An LCA contains interface logic for the micro-controller, memory, and system backplane.

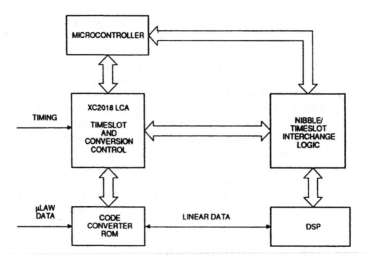

A second LCA implements the glue logic for the data compression circuit.

Figure 1: FPGAs in a voice compression system can be reconfigured to implement internal system diagnostics.

For example, designers at a telecommunications equipment supplier used this strategy in a voice compression module for a T1 multiplexer (Figure 1).[1] The design includes two XC2018 FPGAs. During normal operation, one FPGA provides all the interface logic for the board's microcontroller, RAM, and system backplane, arbitrating accesses to the RAM from the controller and the main system. The second LCA contains most of the 'glue logic' for the data compression operation. However, both FPGAs can be loaded with special diagnostic configurations. In the test mode, the first FPGA connects the microcontroller to the RAM for memory testing, and monitors controls on the system's backplane. The second FPGA can receive timing information from the microcontroller instead of the system backplane, verify the data paths, and check the contents of the EPROM used to implement the code converter's companding algorithm. Actually, two different test configurations have been generated, and others are planned for future product upgrades. All the configuration data is present in memory on the board; the microcontroller handles the downloading of FPGA configuration programs. Similar schemes have been employed in applications such as computer peripherals, industrial controllers, medical equipment, and IC testers. [2, 3]

Often, the quality and reliability of a system can be enhanced by the ability to easily modify units already operating in the field. The alterations might involve fixing a recently-discovered problem or adding new capabilities to an existing design. For systems incorporating SRAM-based FPGAs, such changes are conveniently implemented just by changing the FPGAs' configuration programs. For systems with disk storage, such a logic change can be handled like a software update.

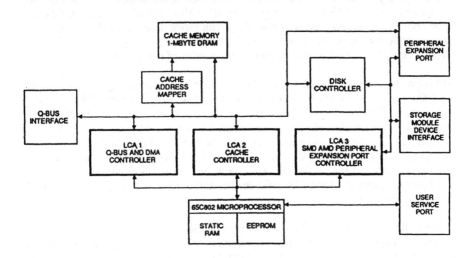

Figure 2: Configuration programs for the FPGAs in this disk controller can be modified through a modem port for loading diagnostics and field upgrades.

This concept was taken one step further in the design of the disk controller shown in Figure 2. [4] Three FPGA devices were used to implement interfaces to the disk, processor bus, expansion port, and cache memory. The configuration programs for the FPGA devices are stored in EEPROM that can be altered using a service port that connects directly to terminals or modems. Diagnostic configurations can be downloaded via this port, allowing for verification of systems in the field directly from the manufacturer's location. In fact, the manufacturer can even modify the system's logic for the customer, such as adjusting the caching algorithm to match the requirements of a particular user.

Boundary Scan Testing

One specific type of diagnostic configuration that can be incorporated in SRAM-based FPGAs is logic to support boundary scan testing of the system. The "bed-of-nails" has been the traditional method of testing electronic assemblies. This approach has become less appropriate, due to closer pin spacings, surface mount technologies, and multi-layer boards. The IEEE Boundary Scan Standard 1149.1 was developed to facilitate board-level testing of electronic assemblies. This specification defines a standard test logic structure implemented with a four-pin interface between Boundary-Scan-compatible devices. By exercising these signals, the user can serially load commands and data into these devices to control the driving of their outputs and examine the state of their inputs. This technique avoids the need to overdrive device outputs, and reduces the user interface to four pins.

Xilinx's third-generation XC4000 FPGA family includes dedicated, built-in logic to implement IEEE 1149.1 Boundary Scan BYPASS, SAMPLE, and EXTEST operations.[5, 6] When placed in boundary scan mode, three normal user I/O pins become dedicated inputs for these functions. The dedicated on-chip logic includes a 4-bit state machine, an instruction register, and a number of data registers.

Although the XC3000 FPGA family does not contain boundary scan registers, an XC3000 device can be configured to emulate the EXTEST function. This external test mode fulfills one of the primary objectives of boundary scan test - the testing of interconnections between ICs. This emulation consumes a significant amount of the XC3000 FPGA's logic resources, and it is not suggested that boundary scan be incorporated into a working design. However, FPGA configurations dedicated to boundary scan testing could be an option in a system containing multiple configuration programs for an XC3000 device, as described in the prior discussion of built-in diagnostics.

A block diagram of the boundary scan emulation circuit for the XC3000 is shown in Figure 3. A specific boundary scan implementation must be created for each FPGA design, matching to the I/O structure of that application. To implement EXTEST, four pins must be dedicated to the Test Access Port (TAP); due to external interconnection requirements, these pins probably cannot be re-used in the actual end application. Eleven of the FPGA's internal logic blocks are needed to implement the TAP controller, Instruction Register, Bypass Register, Test Data Output (TDO) Buffer, and related logic. The TAP controller is implemented as two linked state machines, each using "one-hot" state encoding, and controlled by the serial data entering the Test Mode Select (TMS) input. Test data is shifted from the Test Data Input (TDI) pin, through either the Instruction, Test Data, or Bypass Register (as determined by the TAP Controller), to the Test Data Output (TDO) pin. The Test Data Register contains as many bits as there are used I/O pins in the application being tested, plus one bit for each distinct three-state output control signal. Implementing this logic requires between .5 and 1.5 logic blocks per used I/O pin, plus one additional logic block for each three-state control. Enough logic blocks are available even in the smallest XC3000 device, the XC3020, to accommodate most system designs.

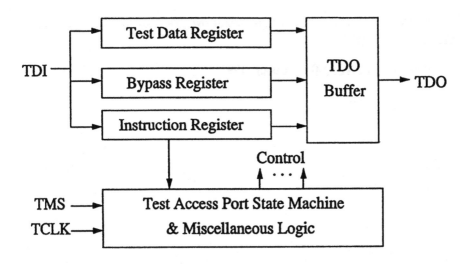

Figure 3: Block diagram of a boundary scan test circuit for the XC3000 FPGA.

Summary

The use of reprogrammable, static-memory-based FPGAs eases system verification in a number of ways. SRAM-based FPGAs provide a fully-tested, high-density device for production use, as well as an efficient prototyping vehicle for verifying custom ASIC designs. Their reconfigurable nature allows for the cost-effective inclusion of self-diagnostic logic in systems, resulting in increased reliability and easier system maintenance. Boundary scan test support is built-in to the XC4000 family architecture, and can be emulated in XC3000 family devices.

References

1. Bradly Fawcett, "Taking Advantage of Reconfigurable Logic", High Performance Systems Programmable Logic Guide, 1989

2. David Smith, "User-Programmable Chips Take on a Broader Range of Applications", VLSI Systems Design, July, 1988

3. Kenneth Hillen, "Build Reconfigurable Peripheral Controllers", Electronic Design, March 8, 1990

4. Jim Reynolds, "Building Tomorrow's Disk Controller Today", Electronic Products, Dec. 15, 1987

5. Jim Donnell, Boundary Scan Puts Tomorrow's Devices to Test", <u>Electronic Design</u>, June 27, 1991

6. Xilinx Inc., <u>The XC4000 Data Book</u>, 1991

MONTAGE: An FPGA for Synchronous and Asynchronous Circuits

Scott Hauck, Gaetano Borriello, Steven Burns, Carl Ebeling

Department of Computer Science and Engineering
University of Washington
Seattle, WA 98195

Abstract. Field-programmable gate arrays are frequently used to implement system interfaces and glue logic. However, there has been little attention given to the special problems of these types of circuits in FPGA architectures. In this paper we describe Montage, a Triptych-based FPGA designed for implementing asynchronous logic and interfacing separately-clocked synchronous circuits. Asynchronous circuits have different requirements than synchronous circuits, which make standard FPGAs unusable for asynchronous applications. At the same time, many asynchronous design methodologies allow components with greatly different performance to be substituted for one another, making a design environment which migrates between FPGA, MPGA, and semi-custom implementations very attractive. Similar problems also exist for interfacing separately-clocked synchronous circuits. We discuss these problems, and demonstrate how the Montage FPGA satisfies the demands of these classes of circuits.

1 Introduction

Field-programmable gate arrays provide an ideal implementation medium for system interface and glue logic. They integrate large amounts of random logic and simple data paths, and can be easily reprogrammed to reflect changes in system components. Unfortunately, most of the effort in designing FPGA architectures has ignored the special problems of these types of circuitry. Interface and glue logic require support for interfacing asynchronous logic to synchronous logic, interconnecting separately-clocked synchronous components, and controlling certain circuit delays [1], all of which are largely ignored by current architectures.

Asynchronous circuits are also not well served by current FPGAs. Implementations of asynchronous logic must consider hazards in the logic, synchronization and arbitration of events, and strict adherence to the timing assumptions of the design methodologies [5, 6]. Unfortunately, it is not possible to implement these circuits in a robust manner in current FPGAs. Some of the elements required (most importantly, arbiters that resolve conflicts between two concurrently arriving signals) are not implementable in the standard digital logic found in these devices. In addition, the logic and routing elements must be designed more carefully in order to avoid extra "glitches" on lines, since in asynchronous circuits every transition is important. Finally, routing resources must have predictable, optimizeable delays in order to meet timing assumptions in the design methodologies.

However, the problems are not restricted simply to the architectures themselves. The mapping software also must be altered to handle the demands of asynchronous logic. Primarily, there are much stricter timing demands that must be upheld. Bundled data and isochronic forks both require signals to be routed with special delay demands, demands that impact placement of logic cells as well. Also, the logic decomposition used to break logic blocks down to the size required by an FPGA (the covering problem) cannot simply use the algorithms for synchronous logic. For quasi-delay insensitive circuits, where the only timing assumptions made are those of isochronic forks, standard synchronous logic decomposition techniques can add extra levels of logic incorrectly, putting hazards into the circuit.

Related work [2] has looked at mapping asynchronous circuits to Actel, antifuse-based, programmable gate arrays. Although the paper outlined an implementation strategy based on a fairly rich library of macro blocks, the underlying limitations of the Actel parts made it difficult to handle arbitration and synchronization, and the complex routing structure did not allow adequate control of routing delays. In addition, we feel that any antifuse-based FPGA is less desirable for asynchronous circuits because of the strict assumptions that must be made about circuit delays. These may require chip delay testing, which is impossible in an unprogrammed antifuse system as path delays cannot be measured until after programming.

2 The Architecture

The Montage FPGA is a version of the Triptych architecture designed to handle synchronous interface and asynchronous circuits. Since much of Montage is identical to Triptych, we direct readers unfamiliar with this architecture to [3]. Like Triptych, Montage is an electrically reconfigurable FPGA, which is preferable to antifuse-based FPGAs because it allows the chip to be programmed for delay testing without permanently configuring it.

The Montage global routing structure is identical to the Triptych routing structure, with diagonal connections between local cells, augmented with vertical segmented channels (see Fig. 1). This structure has proven to be effective for mapping general synchronous circuits. It is even better suited to asynchronous circuits, where one expects to find much more tightly connected subcircuits, and in general less random global routing. Also shared with Triptych is the general philosophy of allowing mappings to fix the tradeoff between logic and routing resources by having logic blocks capable of performing routing functions.

A Montage RLB (shown in Fig. 2) has three inputs and three outputs, and a functional unit (FU) which operates on the three inputs. There are two different types of functional units. The first is a logic block, which implements logic functions and stateholding elements. As shown in Fig. 3, the logic block has a

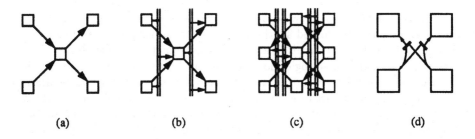

Fig. 1. The overall structure of the Montage FPGA shown in a progression of steps. The basic fanin/fanout structure (a) is augmented with segmented routing channels (b) attached to a third RLB input and output. The structure (c) is obtained by merging two copies of (b), with data flowing in opposite directions in the two copies. Shown in (d) are the connections between the two copies, which permit communication between the two copies, in order to implement sequential logic.

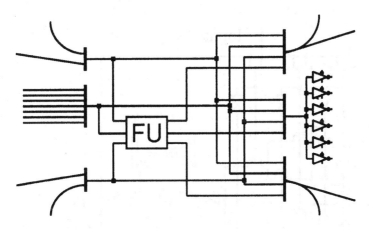

Fig. 2. Montage routing and logic block (RLB) design. The RLB consists of: three multiplexers for the inputs, a functional unit, three multiplexers for the outputs, and tristate drivers for the segmented channels.

function block capable of implementing any function of three inputs. The switch logic function block shown was chosen because it does not suffer from charge sharing. This is important because asynchronous circuits require very clean signals, with absolutely no extraneous transitions. The function output can be fed through a master-slave d-flipflop. This flipflop can be configured with one of two clocks in synchronous mode (allowing two independently clocked synchronous circuits to coexist on a chip), or with a choice of initialization state in asynchronous mode. In the asynchronous initialization mode, the flipflop is set to a value during programming and holds the function output to this value until enough time has passed for the circuit to reach a valid operating state, at which point the two latches comprising the flipflop are made transparent. Each RLB can choose independently how to use the d-flipflop, so on a single chip there can be two separately clocked synchronous circuits, asynchronous elements initialized with the built-in circuitry, and strictly combinational logic blocks. Note that any one of the three logic block inputs can be replaced with a feedback line carrying the flipflop's output value. This feature is used in two separate ways. For synchronous circuit elements, this line carries the flipflop output. This allows a function to be dependent on its previous value, implementing simple state machines in a single cell. For asynchronous elements, this line allows state-holding (non-combinational) elements to be built. This is done by expressing the state-holding function of n inputs as a combinational function of $n + 1$ inputs, where the extra input is the function's previous value. Thus, a single logic block is capable of implementing any three-input combinational function, or a two-input stateholding function such as an asynchronous S-R flipflop or a Muller C-Element.

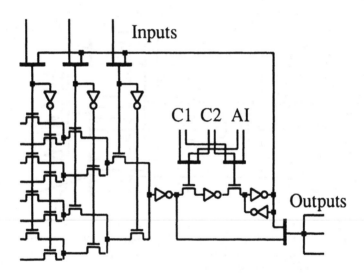

Fig. 3. This logic block is composed of a three input function block and a d-flipflop. The d-flipflop can be put in four modes: transparent (for combinational elements), transparent after initialization (for asynchronous state-holding elements), edge-triggered on clock $C1$, and edge-triggered on clock $C2$.

The second type of functional unit is an arbiter block, shown in Fig. 4. This block is capable of implementing a mutual exclusion element, an enabled mutual exclusion element, or a synchronizer, with all inputs completely permutable and invertable. Although we expect these blocks to be used infrequently, the roles they serve in asynchronous circuits are essential, and are not implementable in standard digital logic. Thus, they must appear as special, built-in blocks in any FPGA which hopes to implement asynchronous circuits, but which does not allow mappings to program circuits at the transistor level (for an example of an EPGA which might allow sufficient transistor-level programming to implement an arbiter, see [4]). For examples of how both types of blocks are used, please see Fig. 6.

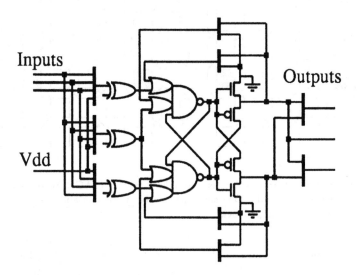

Fig. 4. This arbiter unit can implement either a mutual exclusion element, an enabled mutual exclusion element, or a synchronizer.

Currently we plan to have a 15 : 1 ratio between the number of logic blocks and arbiter blocks. A possible distribution pattern is shown in Fig. 5. This number was chosen based on the relative infrequency of arbiters and synchronizers in typical asynchronous circuits. Since we found that typical Triptych mappings used 25% of their RLBs for routing only, jobs which the arbiter RLBs in Montage are capable of handling, we believe that most unused arbiters will be absorbed into this factor. However, we have taken care to ensure arbiter blocks occupy the same amount of area as logic blocks, allowing easy alteration of the arbiter mix in Montage implementations should it prove necessary.

A last important point to be made about the architecture is how the Montage routing structure handles bundled data and isochronic forks. For bundled data, the simplicity of the Montage routing structure makes it much more feasible to design a router which ensures that bundled-data control signals take longer paths than all of their data bits. See the example shown in Fig. 7. For isochronic forks, there are two different implementation styles dependent on the type of isochronic forks. Isochronic forks can be broken into two classes: one-way, where the only requirement is that a signal reach one end of the fork before it reaches the other; and two-way, where the signal is required to reach both ends of the fork at the same time. For one-way isochronic forks, the signal is routed to the RLB

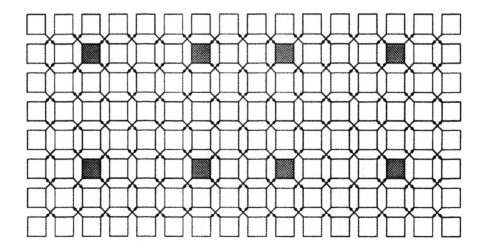

Fig. 5. Distribution of arbiter blocks throughout a MONTAGE array. The shaded blocks represent arbiter units. One sixteenth of the RLBs contain arbiter units. The remaining RLBs contain logic blocks.

of the critical end of the fork, and is then routed back out of this block to the other side of the fork. Thus, the dual routing and logic nature of a Montage RLB ensures that the signal reaches one cell before the other. For two-way isochronic forks, the two ends of the fork are placed either off the same interconnect line, or off diagonals flowing from a shared source RLB. In this way, the isochronic fork depends on the correct speed of very localized elements, delays which can easily be checked during initial chip verification.

3   Future Work

The development of an FPGA for asynchronous circuits opens up several new avenues of exploration. The entire process of mapping for FPGAs must be re-evaluated for this domain. Most obviously, placement algorithms must take into account the constraints generated by bundled data and isochronic forks, and routers must ensure these constraints are met. A more subtle issue arises in the covering process. Quasi-delay insensitive circuits, which make no timing assumptions other than that of isochronic forks, cannot always be decomposed using standard logic decomposition techniques. The problem is that when a logic function is decomposed, the lines introduced to connect the subcomponents may now have transitions on them that did not exist in the original circuit. Since these transitions are obviously not properly sensed in the original circuit, where they did not even exist, it can cause the circuit to malfunction. Thus, new covering techniques must be explored, such as a complete resynthesis from the original circuit description.

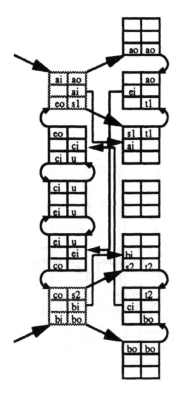
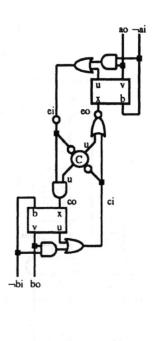

Fig. 6. First example circuit: Martin's fair arbiter [5], built with two synchronizers (arbiter blocks have grey outlines).

Acknowledgments

This research was funded in part by the Defense Advanced Research Projects Agency under Contract N00014-J-91-4041. Gaetano Borriello and Carl Ebeling were supported in part by NSF Presidential Young Investigator Awards, with matching funds provided by IBM Corporation and Sun Microsystems.

References

1. G. Borriello. *New Interface Specification Methodology and its Application to Transducer Synthesis.* P.h.D. thesis, University of California, Berkeley, May 1988. UCB/CSD 88/430.
2. E. Brunvand. Implementing self-timed systems with FPGAs. In *International Workshop on Field-Programmable Logic and Applications*, Oxford, 1991.
3. S. Hauck, G. Borriello, and C. Ebeling. Triptych: An FPGA architecture with integrated logic and routing. In *Brown/MIT Conference on Advanced Research in VLSI and Parallel Systems*, March 1992.

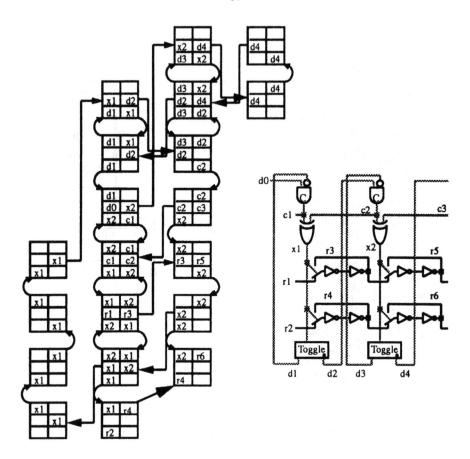

Fig. 7. Second example circuit: Sutherland's micropipelined FIFO [6]. Note that although only two levels of the FIFO are shown, the mapping fits together for longer FIFOs.

4. D. Marple and L. Cooke. An MPGA compatible FPGA architecture. In *First International ACM/SIGDA Workshop on Field-Programmable Gate Arrays*, Berkeley, 1992.
5. A. Martin. Programming in VLSI: From communicating processes to delay-insensitive circuits. In C. Hoare, editor, *UT Year of Programming Institute on Concurrent Programming*. Addison-Wesley, Reading, MA, 1990.
6. I. Sutherland. Micropipelines. *Communications of the ACM*, 32(6), June 1989.

ORCA: A New Architecture for High-Performance FPGAs

D. Hill[1,+], B. Britton[2], B. Oswald[2], N-S Woo[1], S. Singh[2], C-T Chen[2], B. Krambeck[2]

(1) AT&T Bell Laboratories, Murray Hill, New Jersey, U.S.A.
(2) AT&T Bell Laboratories Allentown, Pennsylvania, U.S.A.

Abstract

AT&T's ORCA (Optimized Reconfigurable Cell Array) architecture extends FPGA applicability into a larger domain than is possible with today's parts, including datapath intensive designs such as memory controllers, signal processing parts, and telecommunication interfaces. Key to the suitability of the ORCA for these jobs is the fact that each of its basic blocks is capable of processing four bits. So, for example, a 16 bit adder requires exactly 4 blocks, not 9 or 16 as in other architectures. Yet the total complexity of each block is comparable to other current parts, thus yielding a significant improvement in functional density.

1. Introduction and Background

Early FPGAs (e.g., Xilinx/AT&T 3000 series [10, 11]) have been used mainly to implement random logic circuits. Recent trend of FPGA architecture, however, is to support efficient implementation of data path circuits (as well as random logic circuits). For instance, XC 4000 series FPGAs [12] provide fast 2-bit addition at each logic cell by a dedicated carry circuit. The ORCA FPGA we describe in this paper represents a big step along this direction of FPGA architecture evolution.

While existing FPGA architectures have proven to be successful at supporting small-to-medium control and interface applications, the ORCA architecture is targeted at a larger audience. Specifically, our objectives include:

- Support for larger data manipulations, such as 32 addition/subtraction or normalization.

- Support for wide data registers.

- Support for flexible, on-chip application memories, especially for telecommunication applications.

- Support for systems that mix FPGAs and mask programmed components. It is our objective that a user should be able to switch from one to another painlessly, with minimal engineering time or delay, and with no disruption in functionality.

Following the above goal, we designed the ORCA FPGA to support efficient implementation of data path circuits as well as control (random logic) circuits. The

---

+ D. Hill is currently at Synopsys, Inc.

ORCA FPGA is a SRAM-based FPGA, and it is reprogrammable. Salient features of ORCA architecture include :

1) One logic cell can perform arithmetic operations on a pair of 4-bit data,
2) One logic cell can perform 4-bit binary up, down, or up/down counting,
3) One logic cell can implement a 16x4 memory (RAM or ROM),
4) One logic cell can implement a 4-bit register,
5) One logic cell can implement many combinations of Boolean functions,
6) Routing resource supports 4-bit wide bus interconnection between logic cells,
7) Routing resource is structured in a hierarchical way and, so, it provides flexible routing at low cost.

Typically, the above features 1, 3 and 5 can be combined with 4. (For more details, see the next section.)

There are two types of cells in an ORCA FPGA. One is Programmable Logic Cells (PLCs), which contain both logic and routing resource that can be configured to implement application circuits. The other is Programmable IO Cells (PICs) that are at the periphery of a chip; these cells relay signals between pads and PLCs. Although an ORCA chip contains other function blocks (e.g., configuration circuit and boundary scan logic), we will concentrate on the above two types of cells. Note that only PLCs contain logic resource. This homogeneous architecture makes it much easier to migrate implementations from FPGA to other technologies (e.g., standard cell).

2. Programmable Logic Clusters (PLCs)

As mentioned above, PLC contains both logic and routing resource. The logic resource of PLC is called Programmable Function Unit (PFU). Fig. 1 shows a block diagram of PLC. The *connection switch* (sometimes called *switching R-nodes*) connects inputs/outputs of PFU to (programmable) routing wires and to connection switches of neighboring PLCs. Each PLC is connected directly to its four neighbor PLCs. In order to connect two PLCs that are not neighbors, ORCA provides three different types of routing lines.

2.1 The Programmable Function Unit (PFU)

The Programmable Function Unit (PFU) contains 64-bit SRAM for a lookup table and four flipflops for a register. Fig. 2 shows a detailed diagram of PFU. (We do not show some control signals such as clock, clock enable and reset for simplicity.) The lookup table comprises of four small tables (T3–T0), called *quad lookup tables* (or QLUTs), each of which is 16-bit SRAM. Outputs of the lookup table can be fed to internal flipflops (F3–F0) or to output ports of the PFU. Three multiplexers, M1–M3, are used to support 5- or 6-input Boolean functions. There are five output signals from PFU; each output signal may come from any lookup tables or flipflops.

2.1.1 The Lookup Table

The lookup table of a PLC consists of 64 bits of SRAM and a considerable amount of special decode and routing logic. This logic makes the lookup table highly flexible, and allows it to be used in three distinct modes: logic, arithmetic, and memory.

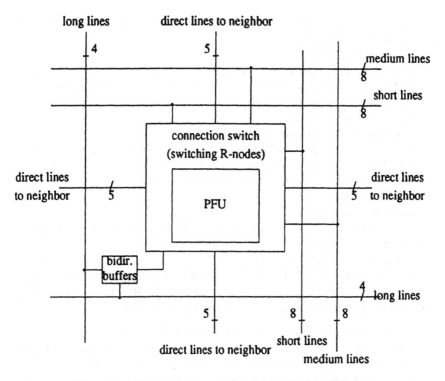

Figure 1. A block diagram of a PLC of the ORCA FPGA.

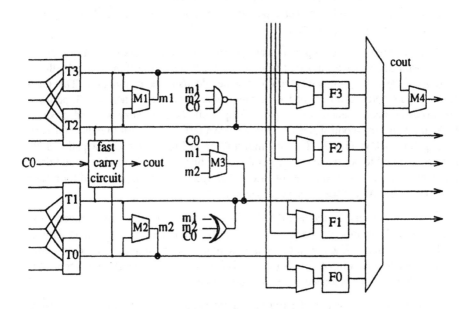

Figure 2. A block diagram of PFU of the ORCA FPGA.

In the logic mode, the 64 bit lookup table is capable of implementing any of the following Boolean functions:

(i) Four 4-input functions.

In this case, each quad lookup table implements one function. Suppose that a function G_i is implemented by QLUT T_i, $0 \leq i \leq 3$. Then, two functions G_j and G_{j+1} ($j = 0, 2$) must share some inputs, i.e., the total number of inputs of G_j and G_{j+1} must be five or less.

(ii) Two independent 5-input functions.

As shown in Fig. 2, the combination of T3, T2 and M1 supports one arbitrary 5-input Boolean function. Similarly, T1, T0 and M2 support another 5-input function. Note that one PFU can implement one 5-input function and two 4-input functions.

(iii) One 6-input function.

The combination of T3, T2, T1, T0, M1, M2 and M3 supports an arbitrary function of six inputs.

(iv) Some functions of up to 11 inputs.

The combination of T3, T2, T1, T0, M1, M2 and 3-input NAND/XOR gates (see Fig. 2) supports some functions of up to 11 inputs. This configuration is useful to implement address recognition and parity circuits.

Experience with this structure has shown that it is efficient for general random logic applications using current techniques in logic synthesis and technology mapping. In particular, the structure can be compared directly with the Xilinx/AT&T 3000 series FPGA [10, 11], and it is almost always possible to map applications from the 3000 series onto the ORCA, with one block from the ORCA replacing two from the 3000 series FPGA, which has 32 bits of lookup table. Yet the availability of the 6-input function and other higher input functions allows users to trade off area for performance on critical paths.

But the size of the lookup table was not selected just to support random logic. It is also well suited for arithmetic and memory functions. In arithmetic mode, the 64 entries can be used to implement a nibble-wide adder/subtracter function. This mode has 8 inputs (plus carry-in) and 4 outputs (plus carry-out). Each QLUT performs addition/subtraction function on a pair of input bits (with a carry-in), and produces both sum and carry-control signals. We provide a fast carry circuit to reduce the carry propagation delay: Fig. 2. The M4 multiplexer in Fig. 2 provides a path for the carry-out signal to be routed on a general routing line. We also provide an additional efficient connection to carry lines of PLC. This special connection is used for a carry chain (for a big ALU): see the next subsection.

In its memory mode, the 64 SRAM cells are organized as two independent 16x2 user accessible memories. Each memory can be configured as either read-write or read-only memory. Each memory requires four address signals, two data output signals, and two data input signals (optional). Designers can specify initial content of each memory, which is loaded at configuration time.

Special bus facilities in the routing fabric simplify the job of creating larger memories. These include nibble-wide 3-state drivers (i.e., "bidirectional buffers" in

Fig. 1) in each PLC. Deep memories may be created at multiples of 16 entries by stacking up PLCs and enabling only the selected PLCs' driver. Naturally, wider memories may be created at multiples of 4 bits: the routing structure is designed to do this with minimal overhead.

2.1.2 Latches

In keeping with the nibble-wide nature of the ORCA architecture, each PFU includes four latches. The four latches can be configured as edge-triggered or level-sensitive. As shown in Fig. 2, data input of each latch comes from either the output of QLUT or a direct input line. Note that we can select the input of a latch dynamically by a control signal. (This feature of a latch is exploited in the example of Section 4.)

All four latches share a clock line, a clock-enable line and a reset line. The clock line may be configured to have either rising-edge or falling-edge clock. The clock-enable line may be configured to be either active high or low. The reset line may be configured to be either asynchronous or synchronous reset; designers may also specify a reset value of each latch. (In addition to this reset facility of each PFU, there is one global reset line per chip. If activated, the global reset signal clears all internal flipflops of the chip.) The latches can be combined with the lookup table to create a flexible ALU circuit.

2.2 The Routing Structure of PLC

The routing structure of the ORCA architecture is designed to support nibble-wide data transmission. The routing structure also has a hierarchy to provide maximum routing flexibility at reasonable cost.

At the higher level, there are three types of inter-PLC routing segments (or routing-resource nodes) : XL (long) lines, X4 (medium) lines and X1 (short) lines. XL lines run across a chip; these lines are generally used for clock signal(s) or high speed signals. X4 lines connect two PLCs whose distance is 4. That is, X4 lines are broken after every four PLCs; this is accomplished by twisting a group of four X4 lines in each PLC, one of which is broken. X1 lines connect two PLCs whose distance is 1; these lines are broken at every PLC.

As shown in Fig. 1, one PLC has four vertical and four horizontal XL lines; one PLC also has eight horizontal/vertical X4 lines, and eight horizontal/vertical X1 lines. Suppose that we want to route a (clock) net on a vertical XL line to a horizontal XL line. Then, all we need to do is to connect the vertical XL line through a programmable bidirectional buffer (Fig. 1) to a horizontal XL line. Therefore, it is easy to build a clock spine to implement application circuits.

At the lower level, there are four types of intra-PLC routing segments: PFU input, PFU output, connection to bidirectional buffers, and "switching R-nodes" (Fig. 1). The switching R-nodes of a PLC provide a direct connection from PFU output (input) of the PLC to PFU input (output) of its neighbor PLC. This direct connection between PFUs is used to support fast connection between two adjacent PLCs. It is also used to provide a fast carry chain between PFUs. That is, the carry-out line of a PFU is directly connected to carry-in lines of all four neighbor PFUs. This direct connection of carry line reduces carry propagation delay. The rich connection of carry line (from one PFU to its four neighbor PFUs) provides ORCA users (or CAD software) with flexibility in implementing large ALU circuits. (This feature contrasts to XC 4000 FPGA [12], in which special carry line runs vertically.)

The switching R-nodes also connect PFU input/output to X1/X4/XL lines as well as one X1/X4/XL line to another.

3. Programmable IO Cells (PICs)

The perimeter PLCs of an ORCA chip are connected to IO pads through Programmable IO Cells (PICs). One PIC connects one PLC to/from four adjacent IO pads to facilitate a nibble-wide bidirectional bus, or eight adjacent IO pads to implement a byte-wide unidirectional bus.

PIC includes four buffers, each of which can be configured as input, output or bidirectional buffer. The input buffers can be configured individually to receive either TTL or CMOS levels with an optional pull-up or pull-down. The CMOS output buffers can be configured to be one of three speeds: slow, medium and fast.

A block diagram of a PIC is shown in Fig. 3. In addition to the (programmable) routing lines connected to a PLC, a PIC also has some routing resource connecting it to other PICs. Note that there is no latches inside PIC. (If a user wants to latch input signals, s/he can use latches in a PLC connected to a PIC.)

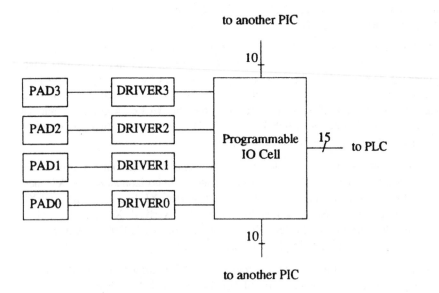

Figure 3. A Programmable IO Cell.

4. An Application Example

Suppose that we want to implement a circuit shown in Fig. 4(a) using an ORCA FPGA. There are two 4-to-1 multiplexers whose outputs are connected to two flipflops. (This circuit is a segment of the PREP benchmark number 1 [14].)

Fig. 4(b) shows an ORCA implementation. In this implementation, each 4-to-1 multiplexer is divided into three 2-to-1 multiplexers; two of them are implemented by QLUTs and the remaining one is implemented by 2-input selection feature of internal flipflop. As a result, we need only one PLC to implement the application circuit.

A direct implementation of a 4-to-1 multiplexer would require 64-bit lookup table, because the multiplexer has six input signals (including two selection signals). Thus if one implements the two multiplexers directly, s/he would need two PLCs.

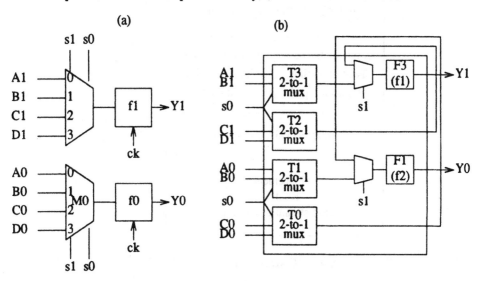

Figure 4. An example of implementing application circuits with ORCA.

5. A Design environment for ORCA FPGAs

A set of CAD tools has been built to design, evaluate, and use the ORCA FPGA [4,5,6,7]. In addition, a set of software tools has been developed to help users implement their application circuits in ORCA FPGAs. These tools include logic optimization, technology mapping, placement, routing and bit-stream generation [13].

Fig. 5 shows layers of CAD tools for ORCA FPGA. Users may describe their application circuits in one of three ways:

(i) using a behavior description language such as Verilog or VHDL,

(ii) using a graphics tool that supports both parameterizable data path elements and finite state machines [8],

(iii) using a schematic capture tool.

In the first two cases, users would use a synthesis tool that exploits special features of ORCA (e.g., [8]).

In any of the above three cases, application circuits will be transformed to netlists of PLCs. If the total number of PLCs needed for a circuit is too large, one may have to go though multiple-FPGA partitioning [9]. Otherwise, one would perform layout by using placement and routing tools.

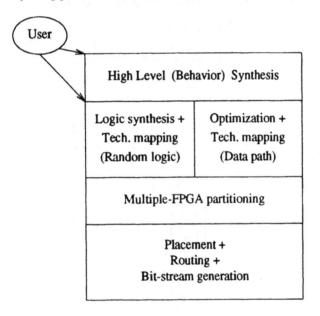

Figure 5. A set of CAD software for ORCA.

6. Summary

The ORCA architecture represents an evolution in the state of FPGA design. It incorporates support for large data paths on a nibble wide basis without diminishing the support for random logic control applications. This is accomplished by providing a function unit which is equally adapted to both environments. The routing structure is similarly designed to allow nibbles of data to be moved around the chip efficiently, while not penalizing the individual connections that characterize typical control logic.

We evaluated the ORCA architecture for both random logic circuits and data path circuits. For random logic examples of the MCNC benchmark, the results after a logic synthesis and technology mapping indicate that ORCA will be somewhat denser than currently available parts, that is, the larger block size is slightly more than offset by fewer blocks being required. For data path circuits, ORCA shows about 2:1 improvement over currently available parts in density, with no degradation in performance.

ORCA FPGA chips are implemented by AT&T Microelectronics 0.8 um, 0.6 um and 0.5 um CMOS processes.

Acknowledgement

We would like to thank to H. Hankinson, C. Lee, W-B Leung, H. Nguyen, Y. Oh, J. Rowland and J. Steward for their help and effort in ORCA implementation.

References

[1] D. Hill, N-S Woo, "The Benefits of Flexibility in Lookup Table-Based FPGAs", IEEE Transaction on CAD, Vol. 12, No. 2, pp. 349-353, February, 1993.

[2] J. Rose, R. Frances, D. Lewis, P. Chow, "Architecture of Field-Programmable Gate Arrays: The Effect of Logic Block Functionality on Area Efficiency," *IEEE Journal of Solid State Circuits*, Vol. 25, No. 5, pp. 1217-1225, October 1990.

[3] D. Hill, D. Cassiday, "Preliminary Description of Tabula Rasa: An Electrically Reconfigurable Hardware Engine," *International Conference on Computer and Design* 1990

[4] D. Hill, "A CAD System for the Design of Field Programmable Gate Arrays," *Proc. of the 28th Design Automation Conference*, June, 1991

[5] D. Hill, "A Specialized CAD Tool for Laying out a New FPGA Block," *3rd Physical Design Workshop*, 1991.

[6] N-S Woo, "A Heuristic Method for FPGA Technology Mapping based on the Edge Visibility," *Proc. of the 28th Design Automation Conference*, pp. 248-251, June 1991.

[7] N-S Woo, "ATOM: Technology Mapping of Sequential Circuits for Lookup Table-based FPGAs," AT&T Bell Laboratories, Technical Report, November 1991.

[8] N-S Woo, M. Cantone, "Scuba: A Synthesis System from Module-based Description of Structured Circuits to ORCA FPGA," AT&T Bell Laboratories, Technical Report, September 1992.

[9] N-S Woo, J. Kim, "An Efficient Method of Partitioning Circuits for Multiple-FPGA Implementation," To be presented at the 30th Design Automation Conference, June 1993.

[10] *Xilinx Programmable Gate Array User's Guide*, Xilinx Corp., 1988.

[11] *ATT 3000 Series Field-Programmable gate Arrays*, AT&T Microelectronics, August 1991.

[12] *Technical Data Book: XC 4000 Logic Cell Array Family*, Xilinx, Corp., 1990.

[13] *ODS: ORCA Design System*, AT&T Microelectronics, January 1993.

[14] *PREP Benchmark Examples*, Programmable Electronics Performance Corporation, September, 1992.

Patching Method for Lookup-Table Type FPGA's

Masahiro Fujita[1] and Yuji Kukimoto[2]*

[1] Processor Laboratory, Fujitsu Laboratories Ltd., Kawasaki 211, Japan
[2] Department of EECS, University of California, Berkeley, CA 94720, U.S.A.

Abstract. Field programmable gate arrays (FPGA) make rapid prototyping an easier task, and are useful in many applications due to their growing speed and capacity. In this paper, we present a rectification method for lookup-table type FPGA's. Instead of changing the netlist of a circuit, we only modify the functionality realized by look-up tables and keep the netlist equal so that there will be no change in the delay of the circuit. We formulate the problem using characteristic functions and present a redesign method based on Boolean relation techniques.

1 Introduction

Field programmable gate arrays (FPGA) are an important technology which has recently attracted much attention due to their advantage of rapid prototyping. There has also been increasing interest in using FPGA's for low-volume production of ASIC designs. Not a few papers have been published so far on logic synthesis for FPGA's, for example Chortle by Francis *et al.* [3] and MIS-pga by Murgai *et al.* [8]. However, no work has been reported on rectifying FPGA designs. Since FPGA is especially useful for prototyping, many design modifications could happen when developing a chip. In the initial design stage, even specifications themselves often change. Furthermore there could be many design modifications to remove design errors. The problem we face with during the redesign phase is that even if one modifies only a small part of a design, the circuit structure can be drastically changed due to logic synthesis. In particular a small modification of a netlist leads to a considerable change in the internal structure of the chip through automatic placement and routing process. This implies that the delay of a redesigned circuit is not predictable after design modifications, which often causes serious problems to designers, since designers take into account the timing characteristics of an original circuit when modifying the design.

In this paper we present an automatic method to rectify lookup-table type FPGA designs. Instead of changing a netlist, we only modify the functionality realized by look-up tables in a chip and retain the netlist so that there will be no

---

* This work was done when the second author was at the University of Tokyo, Tokyo 113, Japan

change in the delay of the chip. We formulate the problem using characteristic functions and present a redesign technique based on Boolean relations.

There has been only limited work on redesign problems. Redesign methods utilizing existing circuits are reported in [4, 5]. [12] discusses another technique for engineering changes. However, since both methods involve the change of circuit structures, these approaches are not so suitable for redesign of FPGA chips.

This paper is organized as follows. In section 2, the proposed redesign technique is informally discussed using examples. Terminology is briefly described in Section 3. Section 4 gives an algorithm to solve the redesign problem. In section 5 experimental results are shown. Section 6 concludes the paper.

2 Example

We give a simple example to make clear what we want to do. A circuit to be rectified is shown in Figure 1(a). The circuit, a one-bit adder, computes the overflow (output o) and the sum (output s) of inputs a, b, and c. Here we assume that each lookup table(LUT) can realize any logic function of up to two inputs. (This is introduced for simplicity, and we can handle any number of fan-ins in the same manner). In the circuit, o and s realize:

$$o = ab + bc + ac$$
$$s = \overline{a}\overline{b}c + \overline{a}b\overline{c} + a\overline{b}\overline{c} + abc$$

Suppose that we want to change the logic function of output o to:

$$o = \overline{a} + \overline{b} + \overline{c}$$

without changing any circuit structure (topology). What we can do is restricted to modifying the functions of LUT's appropriately. Here we first choose a set of LUT's and then try to modify their functions so that the output realizes the required functionality. This can be done by changing the logic function of the LUT for gate g_7 from OR to NAND as shown in Figure 1(b). In this case, modifying one LUT has successfully rectifies the design. Note that the fan-ins of gate g_7 remain unchanged. (g_1, g_5)

Now suppose that we also want to change the function of output s to:

$$s = a + b + c$$

In this case we cannot accomplish the rectification by modifying only one gate. Instead, we have to change the functions of multiple gates simultaneously. In Figure 1(c) a rectification example is shown, where the functions of g_1, g_5, g_6 and g_8 are changed. Again note that the fan-ins of these gates are preserved except g_6, where the original fan-ins are g_1 and g_4, while the redesigned circuit needs only g_4 as its fan-in. This change, however, does not contradict our assumption since no new connections are needed. In other words, the net from g_1 to g_6 is not removed in the redesigned circuit, though the function of the LUT corresponding to gate g_6 is independent of g_1.

If a set of selected gates is not appropriate, we cannot succeed in rectifying a circuit. In such a case, we retry with another set of gates and continue this process until we get a redesigned circuit or give up the patching.

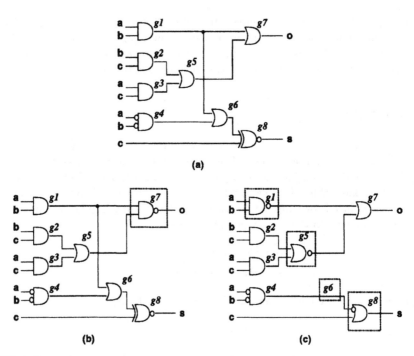

Fig. 1. Examples

3 Terminology

3.1 Boolean Relations

We introduce some definitions to be used throughout the following discussion. The detailed explanation of Boolean relations can be found in [11].

Definition 1. A Boolean relation R is a relation over two sets B^m and B^n, where $B = \{0, 1\}$. $R \subseteq B^m \times B^n$.

Definition 2. A Boolean relation R is *well-defined* if for every minterm $x \in B^m$ there exists a minterm $y \in B^n$ such that $(x, y) \in R$.

Definition 3. For a given Boolean relation R, a Boolean function $f : B^m \rightarrow B^n$ is *compatible* with R if for every minterm $x \in B^m$, $(x, f(x)) \in R$. Otherwise f is *incompatible* with R.

Definition 4. The characteristic function of a Boolean relation R is a single-output function $\mathcal{R} : B^m \times B^n \rightarrow B$ such that $\mathcal{R}(x, y) = 1$ if and only if $(x, y) \in R$.

3.2 Set Operations

The *smoothing operator*, $\mathcal{S}_x f$, is defined as:

$$\mathcal{S}_x f = f_x + f_{\bar{x}}$$

where f_x and $f_{\bar{x}}$ are the cofactors of f with respect to x and $\bar{x}$ respectively.

The *consensus operator*, $\mathcal{C}_x f$, is defined as:

$$\mathcal{C}_x f = f_x f_{\bar{x}}$$

These operations can be performed efficiently using BDD's [1] as a representation of Boolean functions.

4 Rectification Algorithm

In this section we discuss an algorithm to solve the rectification problem addressed in Section 2. The proposed algorithm is formulated based on characteristic functions. Our approach is summarized as follows.

1. Select a set of nodes as a subcircuit where modification is allowed.
2. Compute the Boolean relation permissible for the subcircuit.
3. Restrict the relation by incorporating constraints on supports.
4. Extract from the relation a function which satisfies the constraints.

4.1 Deriving Sets of Candidate LUT's

Firstly a set of lookup tables should be chosen. It is assumed that we can change only the functionality of each table in the set. To select the set of candidate lookup tables, a given network is partitioned into clusters of lookup tables based on topological levelization. The levelization assigns a level to each table in the network from primary inputs to primary outputs or vice versa. The nodes in the same level form a single cluster. A set of candidate lookup tables is selected from these clusters. If redesign turns out to be impossible for a cluster, then another one is tried. The selection of LUT's can be done in various ways.

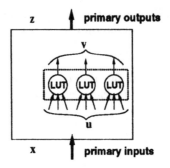

Fig. 2. Multi-level Network

4.2 Permissible Functionality of Subcircuits

Given a set of lookup tables, the permissible functionality of the subcircuit to realize a given specification is computed as a Boolean relation based on Cerny's technique [2]. Cerny's method is formulated with characteristic functions and can be efficiently implemented by BDD's [1]. This topic has recently been revisited in [9, 7]. Here we follow the algorithm presented in [2].

In Figure 2 a multi-level network is shown. The area surrounded by a dotted line represents a subcircuit composed of several lookup tables. In the following, $x = \{x_1, \ldots, x_m\}, z = \{z_1, \ldots, z_n\}$ denote the primary input vector and the primary output vector of the entire circuit respectively. $u = \{u_1, \ldots, u_p\}$ and $v = \{v_1, \ldots, v_q\}$ represent the input vector and the output vector of the subnetwork respectively.

Let $f_i(x)$ be the function realized at u_i in terms of primary inputs. $FI(x)$ is the characteristic function which represents the controllability of u from primary inputs x.

$$FI(x, u) = \prod_{i=1}^{p}(u_i \equiv f_i(x))$$

Let $g_j(x, v)$ be the function realized at z_j in terms of primary inputs x and internal nodes v. $FO(x, v, z)$ is the characteristic function which represents the observability of x and v at z.

$$FO(x, v, z) = \prod_{j=1}^{n}(z_j \equiv g_j(x, v))$$

Theorem 5. (Cerny [2]) *Let $Spec(x, z)$ be the characteristic function which represents a new specification of a given network. To change the functionality of the circuit to $Spec(x, z)$, the output function of a subcircuit must be compatible with Boolean relation $R(u, v)$ defined by*

$$R(u, v) = \mathcal{C}_x(\overline{FI(x, u)} + \mathcal{S}_z FO(x, v, z) Spec(x, z))$$

(See Theorem 3 in [2] for details.)

The derived Boolean relation captures all the functionality permissible for the subnetwork to perform the given rectification. However, since we pose the constraint that each node in the subnetwork has to preserve the support of its function, a further restriction is needed for this relation.

In the remaining part of this paper, we refer to this restriction on the supports of functions as *support constraints*. The Boolean relation of a subnetwork computed based on Theorem 5 is referred to as an *original Boolean relation* and is denoted by $R(u, v)$ or simply R, while the Boolean relation after considering support constraints is called a constrained Boolean relation and denoted by $R_c(u, v)$ or R_c.

```
      restrict_relation(R)
1        do {
2          R_old = R;
3          for each fanin u_i ∈ u
4            V_s = {v_j ∈ v | u_i is an element of v_j's support};
5            R = R · C_{u_i} S_{V_s} R;
6          } while ( R ≠ R_old );
7        return(R);
```

Fig. 3. Restriction Algorithm for Support Constraints

Figure 3 shows an algorithm to perform restriction on an original Boolean relation with respect to support constraints. The procedure receives an original Boolean relation R and returns a constrained Boolean relation $R_c \subseteq R$ after removing unacceptable elements of R. The inner loop from line 2 to 5 is the core part of the algorithm. In line 4 the nodes whose support includes u_i are enumerated in set V_s. In line 5 by performing a smoothing operation on R with respect to V_s, we extract a relation where output patterns for variables in V_s are simply omitted. The relation is restricted by a consensus operation on u_i and ANDed with the original relation. This loop is repeated until the relation reaches a fixed point. Since the operation in line 5 is monotone decreasing with respect to a set of permissible input-output patterns, it is guaranteed that the iteration eventually comes to an end.

Theorem 6. *If there exists a compatible function f with R such that f meets support constraints, then f is also compatible with R_c.*

From Theorem 6, it is guaranteed that if a constrained Boolean relation is not well-defined, then there is no solution which satisfies support constraints. However, from the following counterexample, the reverse is not true.

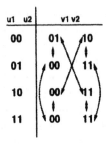

Fig. 4. Boolean Relation

Theorem 7. *Even if a constrained Boolean relation R_c is well-defined, there does not necessarily exist a compatible function f with R_c such that f meets support constraints.*

Proof. We prove this theorem by showing a counterexample. Suppose we get a constrained Boolean relation shown in Figure 4. The relation is clearly well-defined. Here we assume that v_1 depends on only u_1 and v_2 depends on only u_2. One can easily show that the relation has already been restricted by the procedure in Figure 3. Suppose we choose an output minterm 01 for input 00. Since v_1 does not depend on u_2 we have to select a pattern 00 for input 01. Figure 4 shows all these constraints between output minterms of different input minterms. An arrow between two output minterms $x \rightarrow y$ claims that if output minterm x is selected, then output minterm y should be chosen. Figure 4 illustrates that if one chooses a pattern 01 for input minterm 00, then one has to choose a different pattern 10 for the same input minterm 00 and vice versa. For each input minterm only a single output pattern can be selected, so this relation does not have a compatible function which satisfies the support constraints.

The above theorem shows that we cannot handle support constraints exactly with Boolean relations. In Boolean relations the freedom of functionality is represented as a set of permissible output minterms separately for each input minterm. Boolean relations cannot capture our constraints naturally since in the concept of Boolean relations it is assumed that output minterms can be chosen independently for different input minterms.

4.3 Extracting Compatible Functions under Support Constraints

Once a constrained Boolean relation is obtained, we check whether the Boolean relation is well-defined. If not, then from Theorem 6 there is no function which meets support constraints. Otherwise there is possibility that we can complete rectification. Note that we simply need a single solution and are not interested in an optimum solution with respect to some criterion since a lookup table can realize any function of up to n variables in one table no matter how complex it is, where n is a fixed number.

```
       extract_function( R_c )
1        for each v_i in v
2          U_s = {u_j | u_j is not in the support of v_i};
3          R(i) = C_{U_s} S_{V-{v_i}} R_c;
4          if ( R(i) is well-defined )
5            choose a compatible function f_i(u) for v_i;
6            R_c = R_c |_{v_i = f_i(u)};
7          else
8            backtrack to the previous choice point
9        if ( a compatible function is found for each v_i )
10         return( {f_1, ..., f_n} );
11       else
12         return( no function found );
```

Fig. 5. Heuristic Algorithm for Extracting a Compatible Function under Support Constraints

Figure 5 shows a heuristic algorithm which generates from a constrained Boolean relation a compatible function satisfying support constraints[3]. The procedure is invoked if a constrained Boolean relation is well-defined. From Theorem 7, however, the existence of the solution is not guaranteed even if the constrained relation is well-defined. The algorithm is based on a greedy approach, where for each output a function satisfying the constraints is selected(line 5) and the relation is updated by taking this effect into account(line 6). This process is iterated until the functionality of all the output variables is successfully determined. The operations in line 3 extracts the Boolean relation $R(i)$ in terms of u and v_i where the support constraints are satisfied. Since this relation has a single output v_i, it can be easily transformed into an incompletely specified function. For example, for $R(i)$ in line 3, ON-set $= R(i)_{v_i} \overline{R(i)_{\overline{v_i}}}$, OFF-set $= \overline{R(i)_{v_i}} R(i)_{\overline{v_i}}$, and DC-set $= R(i)_{v_i} R(i)_{\overline{v_i}}$. Given these sets, we can easily select a compatible function for v_i in line 5.

Since this procedure involves an exhaustive search in the worst case, we limit the number of backtracks and give up the search if it exceeds a fixed number. Although this approach may overlook a solution, our experiments so far showed that this procedure works well and quickly finds a solution if there exists one since the constrained operation performed on an original Boolean relation prevents us from exploring useless search space.

5 Experimental Results

We have implemented the proposed algorithm on top of SIS[10]. Table 1 shows preliminary experimental results on our rectification method. The MCNC

---

[3] An exact algorithm for this problem can be found in [6].

Table 1. Experimental Results (Sparc Station 2)

circuit	script.rugged followed by FPGA script			only FPGA script		
	result	time(sec)	# of LUTs modified	result	time(sec)	# of LUTs modified
misex1_1	failure	6.3	-	success	1.3	6
misex1_2	failure	6.5	-	failure	6.3	-
misex1_3	success	4.7	5	success	3.7	5
misex1_4	success	4.4	5	success	3.8	5
misex1_5	failure	5.9	-	success	1.3	6
misex1_6	failure	7.2	-	success	3.7	5
misex1_7	failure	5.7	-	failure	5.5	-

benchmark circuit **misex1** is first minimized with **script.rugged** and mapped to Xilinx FPGA's using a script for FPGA's suggested in the manual of SIS. This circuit is assumed to be an original circuit. We obtain new specifications by changing the internal logic of **misex1** intentionally. These modified circuits are named **misex1_1** through **misex1_7**. Finally we check whether the functionality of the original circuitry can be modified to a new specification under support constraints. Out of seven cases we tried, we completed the rectification for two. The right columns of the table contain experimental results where the original circuit is synthesized without **script.rugged**. In this case five specifications out of seven have corresponding redesigned circuits. From the experiments we observed that there are more chances to succeed in redesign before optimization than after optimization. This can be explained as follows. Our approach does not allow any changes in the support of each LUT. Technology-independent optimization in **script.rugged**, however, tries to remove redundant connections to reduce circuit areas. Thus, after optimization, the support of each LUT is likely to consist only of the variables essential to represent its functionality. This suggests that there is much possibility that the support is not enough to change the function to a required one.

6   Conclusions

We have discussed algorithms for rectifying lookup-table type FPGA's under the condition where the support of each LUT is not changed. The constraints we pose enable us to avoid the iteration of expensive placement and routing since the physical layout of an original circuit can be used directly for a redesigned circuit. We have formulated the problem using characteristic functions and have shown that the problem is characterized by Boolean relations and associated constraints, for which we proposed an algorithm to solve.

References

1. R. E. Bryant. Graph-based algorithms for Boolean function manipulation. *IEEE Transactions on Computers*, C-35(8):677–691, August 1986.

2. E. Cerny and M. A. Marin. An approach to unified methodology of combinational switching circuits. *IEEE Transactions on Computers*, C-26(8):745–756, August 1977.

3. R. Francis, J. Rose, and Z. Vranesic. Chortle-crf: Fast technology mapping for lookup table-based FPGAs. In *Proceedings of 28th ACM/IEEE Design Automation Conference*, pages 227–233, June 1991.

4. M. Fujita, T. Kakuda, and Y. Matsunaga. Redesign and automatic error correction of combinational circuits. In *Proceedings of the IFIP TC10/WG10.5 Workshop on Logic and Architecture Synthesis*, pages 253–262. North Holland, May 1990.

5. M. Fujita, Y. Tamiya, Y. Kukimoto, and K.-C. Chen. Application of Boolean unification to combinational logic synthesis. In *Proceedings of IEEE International Conference on Computer-Aided Design*, pages 510–513, November 1991.

6. Y. Kukimoto and M. Fujita. Rectification method for lookup-table type FPGA's. In *Proceedings of IEEE/ACM International Conference on Computer-Aided Design*, pages 54–61, November 1992.

7. Y. Kukimoto and M. Fujita. Reduction of critical path delay by optimizing Boolean relations. In *Proceedings of ACM International Workshop on Timing Issues in the Specification and Synthesis of Digital Systems: Tau92*, March 1992.

8. R. Murgai, N. Shenoy, R. K. Brayton, and A. Sangiovanni-Vincentelli. Improved logic synthesis algorithms for table look up architectures. In *Proceedings of IEEE International Conference on Computer-Aided Design*, pages 564–567, November 1991.

9. H. Savoj and R. K. Brayton. Observability relations and observability don't cares. In *Proceedings of IEEE International Conference on Computer-Aided Design*, pages 518–521, November 1991.

10. E. M. Sentovich, K. J. Singh, C. Moon, H. Savoj, R. K. Brayton, and A. Sangiovanni-Vincentelli. Sequential circuit design using synthesis and optimization. In *Proceedings of IEEE International Conference on Computer Design*, pages 328–333, October 1992.

11. Y. Watanabe and R. K. Brayton. Heuristic minimization of multiple-valued relations. In *Proceedings of IEEE International Conference on Computer-Aided Design*, November 1991.

12. Y. Watanabe and R. K. Brayton. Incremental synthesis for engineering changes. In *Proceedings of IEEE International Conference on Computer Design*, pages 40–43, October 1991.

Automatic One-hot Re-encoding for FPGAs

Dave Allen

Viewlogic Systems, Inc

Abstract. The most common design migrated from PLDs to FPGAs is a state machine. Because of the wide input gates available in PLDs, fully encoded state machines are usually used. However, in register rich FPGAs with narrower gates, one-hot state machines are usually preferred. This paper describes a logic synthesis algorithm which automatically translates a functional level encoded state machine to an equivalent one-hot machine. The result is that without any manual redesign, a PLD state machine can be optimally re-implemented in an FPGA technology such as Xilinx or Actel.

1 Motivation

In PLD state machines, the number of flipflop macrocells is strictly limited and each register input may have a wide fanin. To implement a machine with n states, some binary encoding is chosen to use a minimum number of flipflops, perhaps $k = \log2 (N)$. If there are i inputs and o outputs, then $k + o$ macrocells are used and each of these equations has a fanin of roughly $k + i$. With four inputs, outputs and state bits, each of the eight equations would have roughly eight inputs.

While the encoded technique produces optimal results for PLDs, the constraints for FPGAs are different. Many registers are available, and equations with high fanin are expensive in terms of area and speed. Several major FPGA vendors recommend a technique known as one-hot encoding [1,2]. Each state is represented by a single register. The fanin to the flipflop is limited to the number of transitions into its state. The number of transitions to each state is typically a small constant; for example, each state in a counter has but a single transition into it.

The remainder of this paper is divided into three sections. In the first section, the concept for the automatic translator is presented. In the second section, implementation details are discussed. The final section shows results for a selection of example circuits.

2 Concept

Figure 1 shows how a fully encoded state machine can be re-encoded. At the top is shown a general encoded machine with k state bits and n valid states. The k state bits can be replaced by n one-hot state bits, as shown in the center of Figure 1. On the input side of the new state vector, we place a $k{:}n$ decoder; based on the encoded inputs, only one of the n output bits is active. On the output side of the registers, we place a $n{:}k$ encoder;

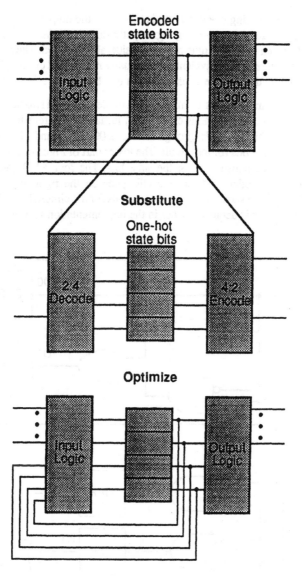

Figure 1. Concept of translator

the binary index of the single active input is placed on the outputs. The resulting logic is quite inefficient. However, by using logic synthesis and the fact that only one of the state bits is active at a time, the logic can be considerably reduced. The result, as shown at the bottom of Figure 1, is equivalent to a manually redesigned circuit. The remainder of this section works through the translation of a two bit counter.

The upper part of Figure 2 shows a simple circuit implementing a two bit counter. Either power on reset, or an asynchronous clear input, initializes the registers to zero. Each clock causes the machine to cycle through the states 00, 01, 10 and 11. On the right are shown the logical equations for the circuit. The lower part of Figure 2 shows the section of logic which will be substituted for the state bits. The four AND gates on the left implement the 2:4 decoder, and the two OR gates on the right implement the 4:2 encoder. On the right are shown the logical equations for the circuit. Note that the output of bit 0 is not used; this is detailed further in the implementation section.

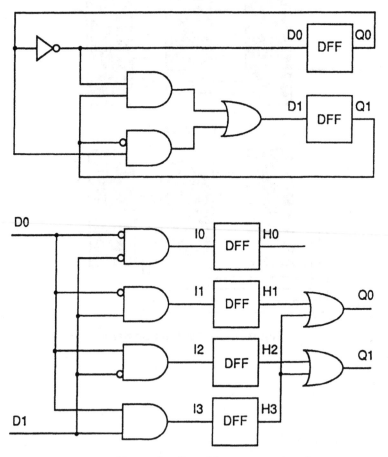

Figure 2a. Two bit counter circuit

The next paragraphs work through the logic simplification and show two surprising tricks which allow the efficient implementation of the equations.

Rewriting D0 and D1,

D0 = NOT (H1 OR H3) = (NOT H1) AND (NOT H3)

D1 = ((H1 OR H3) AND NOT (H2 OR H3)) OR (NOT (H1 AND H3) AND (H2 OR H3))

The equation for D1 shows the first surprising simplification. Since we know that only one of the H bits can be active at once, we can write

For all i not equal to j, Hi AND NOT Hj = Hi

If one state bit is high, we already know all the other bits will be low; therefore, any active low product terms are redundant. Thus the equation for D1 can be reduced to

D1 = H1 OR H2

Continuing,

I1 = (NOT H1) AND (NOT H3) AND NOT (H1 OR H2)

I2 = (H1 OR H3) AND (H1 OR H2)

I3 = (NOT H1) AND (NOT H3) AND (H1 OR H2) = H2

Now for the second surprising simplification.

For all i not equal to j, Hi AND Hj = 0

The situation where two state bits are both active is impossible. This simplifies the equation for I2:

I2 = H1

Summarizing,

I1 = (NOT H1) AND (NOT H2) AND (NOT H3)

D0 = NOT Q0	Q0 = H1 OR H3
D1 = (Q1 AND NOT Q0) OR (Q0 AND NOT Q1)	Q1 = H2 OR H3
	I1 = D0 AND NOT D1
	I2 = D1 AND NOT D0
	I3 = D0 ANDD1

Figure 2b. Two bit counter equations

I2 = H1

I3 = H2

Figure 3 shows the resulting shift register circuit. Note that since asynchronous reset is assumed, the start state is 000. This could be called *no-hot* initialization. As expected, the input to register H1 is active only during this state. The implementation section will explain the algorithm needed to modify this circuit to be truly one-hot.

3 Implementation

This section describes several aspects of the algorithm in more detail. First, state extraction and decoder optimization will be discussed. Second, the method for logic minimization will be covered, and finally the relationship between no-hot and one-hot initialization will be described.

3.1 Decoder optimization

In most state machines, the number of valid states is slightly less than the full 2 to the k possibilities. In a BCD counter, for example, only 10 of the 16 possible states are used. In PLD implementations, these states are either forced to transition to zero, or are used as don't-cares to minimize the number of product terms needed. When unused states exist, this information can be used to simplify the $k{:}n$ decoder. Since unused states do not exist in the one-hot scheme, this is the only place in which don't-care information is relevant.

Although sophisticated algorithms exist for discovering unused states [3], the processing needed to prove whether each of the 2 to the k states is reachable is nontrivial. In fact, even for single PLD size designs, the processing time for this algorithm would dominate the transformation. Since state encoding documentation is available for most PLD designs, even if the source code is unavailable or in an incompatible format, the algorithm in this paper requires manual state extraction. You enter into a file all of the binary encodings corresponding to valid states. Otherwise, the algorithm assumes all states are used.

The valid state information is used to simplify the decoder. As shown in the lower part of Figure 2, the decoder consists of n k-bit AND gates. A two-level don't-care expander, such as Espresso [4], optimally expands the valid states and removes un-needed AND

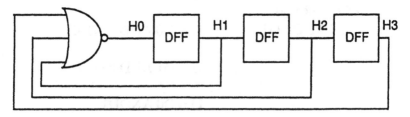

Figure 3. One-hot two bit counter

inputs. For example, if the counter in Figure 2 had three valid states instead of four, it might count the sequence 00, 01, 10. The third state would be expanded to 1- and one AND gate would be removed by the optimization.

3.2 Logic optimization

Figure 2 shows that the original PLD AND-OR structure is replaced with a four level OR-AND-OR-AND structure. Performing the logic optimization "tricks" described earlier is easy once the four level form is converted back to two levels. In this sum of product form, detection of the one-hot simplifications can be done with a single string processing step. The product term is represented by a string of 0,1,- characters in the standard Berkeley PLA or BLIF formats. The '1' character indicates that the input is in the product term; the '0' character indicates that the complement of the input is in the product term; the '-' character indicates that the input is not in the product term. Some of the inputs may not be one-hot signals. However, if more than a single one-hot signal is '1', the entire product term is ignored. Likewise, if a single one-hot input is '1', all '0' one-hot inputs can be changed to '-'. This operation completes the one-hot optimization.

3.3 No-hot initialization

Although many PLDs use 0 as the start state, this is not a requirement. An arbitrary start state may be defined with a mixture of asynchronous sets and clears. If a start state is not specified, the zero state becomes the start state by default. The one-hot register corresponding to the start state must have an asynchronous set, while all the remaining registers must have an asynchronous clear. This ensures that the machine comes up in the correct one-hot state.

An interesting situation arises when the start state is zero. This was briefly mentioned in the example beginning this paper. As that example showed, the zero bit will never be required as an input to the OR gates forming the encoder. One is tempted to optimize away that flipflop, and as the example showed, this will result in a correct implementation. However, in a larger machine, or if the zero bit must be decoded for use in an output, this is not efficient.

When the zero bit is optimized away, a $n-1$ bit wide NOR gate is required to decode that state. Therefore, the algorithm keeps the zero bit as an unused output through the logic optimization described above. After minimization, but before technology mapping, the algorithm searches for this NOR gate and replaces it with the output of the zero bit. This step modifies the design so that instead of having a no-hot initialization, it becomes strictly one-hot.

4 Results

Certain types of state machines may be inappropriate for the algorithm described in this paper. Specifically, if state assignment is done so that a number of state registers are used as primary outputs, this algorithm may not be appropriate. Without going into more detail, it may be worthwhile to apply the algorithm to the state bits which are not primary outputs. The resulting machine could be referred to as "partially one-hot", or "warm". The circuits to which this algorithm was applied were selected with this criterion in mind.

We chose to implement the circuits in the Actel Act-1 technology using a Viewlogic optimization tool. This product accepts Actel netlists or source files in one or more PLD HDLs; it produces optimized netlists. A custom program using the PLA format was written to perform the one-hot transformation and special-purpose optimization. Each circuit was optimized and the delay estimate provided by the optimizer is reported in the table below. Since the processing flow involves several programs communicating with text files, operational speed is dominated by file I/O. Running on a 20 MHz PC-386, no circuit required more than a minute or two of processing to perform the complete transformation. Optimization and technology mapping took between two and ten minutes on the same machine for the circuits shown here.

5 Conclusion

Designers need to automatically translate PLD designs into FPGAs. However, the optimal state machine encoding approaches for the two architectures are quite different. The algorithm described in this paper automatically translates from the fully encoded state machines typically found in PLDs into the one-hot encoding preferred by FPGA vendors and designers. The results were presented for a selection of benchmark designs synthesized to the Actel Act-1 technology. Speedups ranged from 28% to 50%, with area changes ranging from 5% decrease to 27% increase.

References

[1] Alfke, Peter, "Accelerate FPGA Macros with One-hot Approach", Electronic Design, September 13, 1990.

[2] "FPGAs are Better for State Machines than PLDs", FPGA Design Guide, Actel Corporation, August 1991.

[3] Ashar, P, A. Ghosh, S. Devadas, A. R. Newton, "Implicit State Transition Graphs: Applications to Sequential Logic Synthesis and Test", Proc ICCAD 1990, pp 84-87.

[4] Brayton, R. K, G. D. Hachtel, C. T. McMullen and A. L. Sangiovanni-Vincentelli, *Logic Minimization Algorithms for VLSI Synthesis*, Kluver Academic Publishers, 1985

Circuit Name	State Bits	Valid States	Encoded Area			One-hot Area			Encoded Delay	One-hot Delay
			Gates	Flops	Total	Gates	Flops	Total		
Moore	3	4	8	6	14	8	8	16	23.50	18.60
Trafsig	8	11	32	16	48	41	20	61	55.80	37.20
Div10	4	10	9	8	17	0	20	20	18.60	9.30
Dram4B	4	16	81	8	89	54	32	86	65.10	46.50

Table 1. Results

Minimization of Permuted Reed-Muller Trees for Cellular Logic Programmable Gate Arrays

Li-Fei Wu, and Marek A. Perkowski

Department of Electrical Engineering, Portland State University
P.O. Box 751, Portland, Oregon 97207, tel. (503) 725-5411

ABSTRACT

The new family of Field Programmable Gate Arrays, CLI6000 from Concurrent Logic Inc realizes the truly Cellular Logic. It has been mainly designed for the realization of data path architectures. However, introduced by it new universal logic cell calls also for new logic synthesis methods based on regularity of connections. In this paper we present two programs, exact and approximate, for the minimization of Permuted Reed-Muller Trees that are obtained by repetitive application of Davio expansions (Shannon expansions for EXOR gates) in all possible orders of variables in subtrees. Such trees are particularly well matched to both the realization of logic cell and connection structure of the CLI6000 device. It is shown on several standard benchmarks that the heuristic algorithm gives good quality results in much less time than the exact algorithm.

1. Introduction

There is recently an increased interest in logic synthesis using EXOR gates [1,3-14]. New technologies, PLD and FPGA, either include EXOR gates, or allow to realize them in "universal logic modules". Exor circuits have smaller cost than inclusive (AND/OR) circuits. They are always very easily testable and have universal tests [5,13]. One way to realize EXOR-based circuits is through GRMs. The *Generalized Reed-Muller (GRM)* forms, which also include Reed-Muller (RM) form [9], are canonical. All literals in the RM form are positive. In GRM forms all variables are in a fixed form, i.e. each occurrence of a variable in products of the form is either consistently positive or consistently negative. The problem of finding the minimal

This research was partially supported by the NSF grant MIP-9110772}

GRM form of optimal polarity [13] (called also fixed-polarity Reed-Muller), as well as the problem of finding the minimal Exclusive-OR Sum of Products (ESOP) of a Boolean function [1,7,10], are the classical ones in logic synthesis theory. Recently efficient solutions have been proposed: to ESOPs in [1], and to GRMs in [13]. However, to our knowledge, with an exception of [14] and [12] there are no any programs to minimize multi-level (more than three levels) circuits that use EXOR gates.

Recently several companies, including Concurrent Logic [2] and Algotronix brought to market a new generation of FPGAs that can be called "cellular", since the number of global connections for a cell is very limited and most of the connections are only to the abutting cells. CLI chip is a rectangular array of cells [2]. The basic cell of CLI6000 can be programmed to the *two-input multiplexer* $(A * B \oplus C * \overline{B})$ and the *AND/EXOR cell* $(A * B \oplus C)$ as two of its most efficient combinational logic uses. This suggests using these cells for tree-like expansions. The method outlined in this paper is particularly suitable for these new FPGAs for which no special design methods have been yet proposed. These "cellular logic" devices require regular connection patterns in the netlists resulting from logic synthesis.

The AND/OR factorization (MIS II) creates irregular circuits that are hard to map to CLI6000 cells. ESOP factorization approach from [12] creates multi-level circuits that better utilize AND/EXOR cells but are still irregular. This makes their routing to cellular logic devices (specifically CLI6000) extremely difficult and many cells are wasted while programmed as connections only. Several concepts of logic synthesis for cellular logic using EXORs have appeared in past in the literature which may be used to minimal implementations of binary and multiple-valued logic functions [3,8,10]. This paper presents a new tree searching program which generates AND/EXOR tree circuits. The Permuted-Reed-Muller-Tree (PRMT) searching program presented here, *REMIT (REed-Muller Ideally permuted Trees)*, accepts a completely specified Boolean function in the form of an array of ON disjoint cubes as the input [4]. After decomposing the function with respect to all input variables according to *Davio Expansion* the program returns the exact PRM tree (the tree structure that minimizes the cost function) -- in version REMIT-EXACT, or the quasi-minimum PRM tree -- in version REMIT.

2. Davio Expansions

Decomposition is commonly used in logic minimization. Using Shannon and Davio Expansions, the decomposition is achieved step-by-step with respect to all input variables. This kind of decomposition can be applied always, in contrast to other types of Boolean decompositions that are applicable to only some functions, and for which checking the decomposition possibility is very time consuming. The well-known *Shannon Expansion* can be applied to decomposition with an EXOR gate [3]. It is called *Davio expansion*, and is defined as follows:

$$f = x_i \cdot f_{x_i} \oplus \overline{x_i} \cdot f_{\overline{x_i}} \quad (1)$$

$$f = f_{\widetilde{x_i}} \oplus \overline{x_i} \cdot [\, f_{\widetilde{x_i}} \oplus f_{\widetilde{x_i}} \,] \qquad (2)$$

$$f = f_{x_i} \oplus \overline{x_i} \cdot [\, f_{x_i} \oplus f_{\widetilde{x_i}} \,] \qquad (3)$$

where $f_{x_i} = f(x_1,...,x_{i=1},...,x_n)$ and $f_{\widetilde{x_i}} = f(x_1,...,x_{i=0},...,x_n)$. Let us observe that these expansion formulas have been applied by several authors for the synthesis of GRM forms for completely specified functions [11]. Davio [3] and Green [6] use them as a base of *Kronecker Reed-Muller (KRM)*, *Pseudo-Kronecker Reed-Muller (PKRM)*, and *Quasi-Kronecker Reed-Muller (QKRM)* forms (Green uses also trees for better explanation). If only rule 2 is used repeatedly for some fixed order of expansion variables, the RM Trees are created, which correspond to RM forms after their flattening. If for every variable one uses either rule 2 or rule 3, the GRM Trees are created, from which GRMs are obtained by flattening (which proves in other way why there is 2^n of such forms). If for every variable one uses either rule 1, rule 2, or rule 3, the KRM Trees are created, from which KRMs are obtained by flattening (which proves in other way why there is 3^n of such forms).

If rules 1, 2 and 3 are used, but in each subtree there is a choice of a rule, the PKRM Trees are generated from which PKRM forms are obtained by flattening. Now, if additionally we allow the expansion variables to have various orders (but the same in the entire tree), one obtains the QKRM Trees, and QKRM flattened forms, respectively. Allowing various orders of variables in subtrees creates an even wider family of trees [10,11], called *Permuted Trees*. The trees based on expansion (2) but with different orders of variables in subtrees will be called *Permuted Reed-Muller Trees*. They include RM Trees as special cases when the same variable is used for expansion on the same level of the tree. This paper presents an efficient PRM tree searching method.

3. Tree Search to find PRM Trees

For a given Boolean function, after using Davio Expansion to decompose with respect to variable x_i, the function will be divided into two parts: $f_{\widetilde{x_i}}$ and $(f_{x_i} \oplus f_{\widetilde{x_i}})$ or g_{x_i}. Both parts don't contain variable x_i any more. The parts are connected by an EXOR-gate and an AND-gate. The AND-gate takes (x_i) and $(f_{x_i} \oplus f_{\widetilde{x_i}})$ as its two inputs; the EXOR-gate takes $(f_{\widetilde{x_i}})$ and the output of the AND-gate as its two inputs. For each part $f_{\widetilde{x_i}}$ and $(f_{x_i} \oplus f_{\widetilde{x_i}})$, we use again Davio Expansion to decompose for another variable, say x_j. Both parts will be divided into another two parts. At each level of the tree the process is repeated, until all variables are decomposed. The output of the original function f is the root of the tree. At each level there can be different input variables. This kind of decomposition is ideally suited to regular array realization of combinational logic, such as one offered by "cellular logic" FPGAs. In many experiments with multi-output functions we found that the shape of a tree is not a trapezium, as it might be expected, but is closer to a rectangle. There are thus not many cells wasted for routing.

Even though every variable's decomposition is according to the same form (2), the variable decomposing order still makes the results different. The *minimum PRM Tree* is one that has the minimum regular layout-related cost (the weighted cost function takes many factors such as the total number of gates, connections, shape, etc.). The PRM tree searching program introduced here can find the best variable decomposing order and obtain the exact solution for functions of not more than 9 variables. The diagram from Fig. 1 illustrates the search tree created to find the exact PRM Tree. This tree describes all applications of Davio Expansion with respect to all possible variables in all subtrees. (1) Suppose the initial function has n variables. At the first level of the tree (the root of the tree), there is only one node, which stores the cube set of the initial function. (2) At the second level, there are $2n$ nodes. Nodes of each pair correspond to the two parts of the function from Davio Expansion. They are stored as sets of disjoint cubes. (3) To each node (cube set) at the second level Davio Expansion is further applied to decompose other variables as their parent_node did, but the number of decomposing choices is less by one than in the parent_node (because the number of choices is determined by the number of remaining variables in the cube set). (4) At the third level, there are $2n*(n-1)$ nodes. (5) Each node continues the decomposing process until all variables are decomposed (no more variable remains in the cube set). (6) In the worst case, the tree has the maximum of n levels. At the last level there is a total of $n!*2^n$ nodes.

4. Exact PRM Tree Searching Algorithm

The goal of PRM tree searching in REMIT-EXACT is to find the *best variable decomposing order* for PRM circuits. The *best* variable decomposing order will lead to the circuit having the smallest cost. For achieving the best solution, we use full_tree searching. *Full_tree_searching* searches the whole tree to look at all possible combinations of different variable orders in all subtrees independently. Various cost functions can be declared, corresponding to costs of AND, OR, AND/EXOR gates, the numbers of connections, etc. The memory-efficient *backtracking full-tree-searching algorithm* with cut-offs based on costs is used. It generates only the solutions that are not more expensive than those previously generated. The *Decomposing_Tree* searching starts at the root, and the PRM_Tree creating at the leafs. Although backtracking allows to discard some obviously worse solution candidates, still a large tree has to be searched.

5. Heuristic Tree Searching

The heuristic algorithm, REMIT, uses certain heuristics to select the best expansion variable for every subtree of the Decomposing tree. However, at certain levels of the tree all possible variable decomposing orders for each branch are also checked. The variable selection rules are close to those presented in [13,10], for instance one of the heuristics states that the variable that occurs most often in disjoint cubes should be selected. The user has a number of parameters to control the depth and the type of a tree at which the full_searching is executed. For instance, he can request that when the sub-function in the Decomposing tree has 8 variables, the full_searching is executed for it.

6. Evaluation of Results

The test results shows that, regardless how many input variables or how many input cubes, the full_tree searching always returns the best solution (Table 1). I/N denotes the number of inputs, O/N -- the number of outputs, #AND_gate -- the number of AND gates, #EXOR_gate -- the number of EXOR gates. One can see that this method is practical for less than nine input variables.

The heuristic tree searching is designed to solve larger functions. When input function has more than nine variables, the full_tree searching takes extremely long time, because the size of the tree increases exponentially with the number of inputs. The tree searching time increases then also exponentially. For example, for an eight-input function, the full_tree searching takes 200 seconds but a nine-input function may take up to eight hours. Thus, using full_tree searching on high number input functions (more than nine variables) is not practical.

Our heuristic tree searching has rationally solved this problem (Fig. 2). The result quality varies with the different initial functions. For some functions the heuristic tree searching will find the best solutions, while for others the results from the heuristic tree searching are quite different from the results of the full_tree searching. But all results fall obviously in an acceptable region of the best result (Table 2).

A comparison graph from Fig. 2 shows the searching time percentage when using the heuristic tree searching compared with using the full_tree searching. The decrease in searching time is very considerable. The graph from Fig. 3 compares results (cost for the AND-EXOR circuit). It shows the cost increasing percentage of the heuristic tree searching comparing with the full_tree searching. In can be seen from Fig. 3 that the costs given by the heuristic tree search are only slightly higher than the costs given by the full_tree searching.

After testing many functions of a high number of inputs on the heuristic tree searching (with number of inputs varying from 10 to 20), it was found that the speed of finding the results is no longer a problem. Since there are no results comparison with the full_tree searching for functions of many inputs, it is difficult to evaluate exactly how close are the results provided by the heuristic tree searching to the exact minimum results. We have, however, reasons to believe that the results of the heuristic tree searching are good, because it includes many partial full_tree searches.

7. Conclusion

A practical program with exact and quasi-minimum options for the minimization of PRM Trees to be realized in cellular-logic FPGAs has been introduced. It is included in a comprehensive design automation system for the CLI6000 series, that includes special state-machine and physical design programs as well. REMIT gives good results on completely specified functions. Contrary to most papers from the literature, the algorithm uses cubes, not minterms [10,11,13]. Other programs of the system generalize this approach to GRM Trees for incompletely specified multi-output functions [10], and to Kronecker Trees [6,10,11] (which use also multiplexer cells of CLI6000). Similarly as the Shannon expansion is a base of the Binary Decision Diagrams (BDDs), which are a very efficient representation for Boolean functions manipulation, the Davio Expansions are the base of our new *EXOR Decision Diagrams* which exhibit similar properties to BDDs, but reflect the AND/EXOR structure of CLI physical resources. This will be published elsewhere.

The methods shown here open a wide area of interesting and new applications, especially to cellular FPGAs, in which high regularity of connections is more important than the number of logic blocks -- a requirement gives priority to structures such as trees, DAGs, and levelized decompositions with EXOR pre- and post- processors. While for other FPGAs the logic synthesis and the physical design stages are performed separately, for CLi 6000 we advocate a new design approach, which we propose to call *geometrical logic synthesis*. This approach combines the stages of logic synthesis and rough routing/placement into a *single*, comprehensive, and cellular-medium-related, stage.

References

[1]   Ph.V. Besslich, *Proc. Euro ASIC'91*, Paris, 1991.

[2]   Concurrent Logic Inc., "CLI6000 Series Field Programmable Gate Arrays", *Preliminary Information*, Dec. 1, 1991, Rev. 1.3.

[3]   P. Davio, et al, *Discrete and Switching Functions*. George and McGraw-Hill, New York, 1978.

[4]   B.J. Falkowski, I. Schaefer, and M.A. Perkowski, "Effective Computer Methods for the Calculation of Hadamard-Walsh Spectrum for Completely and Incompletely Specified Boolean Functions," *Accepted to IEEE Trans. on CAD.*, 1991.

[5]   H. Fujiwara, *Logic Testing and Design for Testability*, Computer System Series, The MIT Press, 1986.

[6]   D. Green, *Int. J. Electr.*, Vol. 70, No. 2, pp. 259-280, Jan. 91.

[7]   M. Helliwell, and M.A. Perkowski, *Proc. of 25th DAC* 1988, pp. 427 - 432.

[8]   S.L. Hurst, *The Logical Processing of Digital Signals*, Crane-Russak, New York, 1978.

[9]   D.E. Muller, *IRE Trans. Electron. Comp.*, Vol EC-3, pp. 6-12, Sept. 1954.

[10]   M.A. Perkowski, P. Dysko, and B.J. Falkowski, *Proc. IEEE Int. Phoenix Conf. on Comp. and Comm.*, Scottsdale, AZ, pp. 606-613, March 1990.

[11]   M.A. Perkowski, *Proc. of 22-nd ISMVL*, pp. 442-450, Sendai, Japan, May 27-29, 1992.

[12]   M.A. Perkowski, and A. Shahjahan, "Efficient Rectangle Factorization Algorithm for ESOP Circuits", *PSU Report*, Dec. 1991.

[13]   A. Sarabi, and M.A. Perkowski, pp. 30-35, *Proc. DAC'92*, June 1992.

[14]   J.M. Saul, *Proc. IEEE ICCD'91*, Sept. 17-19, 191, pp. 634-637.

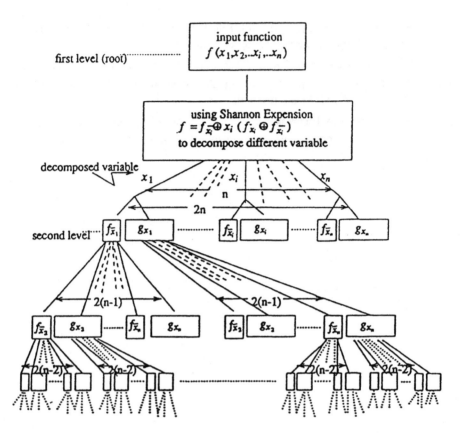

Fig. 1. Diagram for the full_tree searching

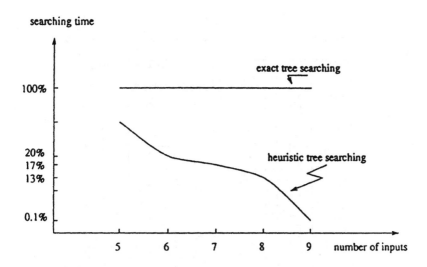

Fig. 2. Heuristic searching time versus exact searching time

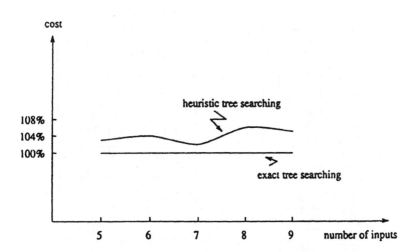

Fig. 3. Cost of the heuristic searching versus the exact searching

TABLE I

GATE NUMBER COMPAROSIN OF EXACT SEARCHING AND RANDOM SEARCHING

Example Name	I/N	O/N	exact-tree searching		random order searching	
			# And-gate	# Exor-gate	# And-gate	#Exor-gate
misex25.tt	6	1	8	11	11	11
misex53.tt	6	1	6	7	10	7
misex57.tt	6	1	8	11	13	11
misex22.tt	6	1	8	11	15	11
misex55.tt	6	1	8	11	12	11
misex58.tt	6	1	6	7	10	7
misex56.tt	6	1	8	11	13	12
misex21.tt	6	1	10	15	15	16
5x6.tt	7	1	3	4	4	4
5x1.tt	7	1	13	15	26	15
5x5.tt	7	1	5	6	7	8
z44.tt	7	1	2	2	2	2
con11.tt	7	1	7	10	9	10
con12.tt	7	1	5	7	5	7
f55.tt	8	1	3	4	4	4
f54.tt	8	1	5	6	8	8
f53.tt	8	1	9	10	17	15
f52.tt	8	1	14	18	33	26
f51.tt	8	1	25	30	50	44

TABLE II

TIME COMPARISON OF EXACT SEARCHING AND HEURISTIC SEARCHING

Example Name	I/N	O/N	exact-tree searching				heuristic searching			
			# And	# Exor	time	level	# And	#Exor	time	level
misex25.tt	6	1	8	11	2.1	6	9	11	0.4	6
misex53.tt	6	1	6	7	1.4	5	8	7	0.3	5
misex57.tt	6	1	8	11	1.9	6	9	11	0.4	6
misex22.tt	6	1	8	11	2.3	6	9	11	0.4	6
misex55.tt	6	1	8	11	2.3	6	10	11	0.5	6
misex58.tt	6	1	6	7	1.8	5	6	7	0.3	5
misex56.tt	6	1	8	11	2.5	6	10	11	0.5	6
misex21.tt	6	1	10	15	3.1	6	10	15	0.6	6
5x6.tt	7	1	3	4	3.0	4	3	4	0.4	4
5x1.tt	7	1	13	15	21.3	7	14	15	3.9	7
5x5.tt	7	1	5	6	7.3	4	5	6	1.3	4
z44.tt	7	1	2	2	1.3	2	2	2	0.1	2
con11.tt	7	1	7	10	8.2	6	8	10	1.2	6
con12.tt	7	1	5	7	7.7	4	5	7	1.4	4
f55.tt	8	1	3	4	24.3	4	4	4	0.2	4
f54.tt	8	1	5	6	45.4	5	5	6	6.1	5
f53.tt	8	1	9	10	102.0	6	10	11	15.4	6
f52.tt	8	1	14	18	243.2	7	16	19	32.6	7
f51.tt	8	1	25	30	426.3	8	26	31	57.1	8
9asy.tt	9	1	41	73	13760.0	9	48	73	110.2	9
misex64.tt	10	1					147	295	321.6	10
misex47.tt	11	1					53	39	573.2	11
misex60.tt	12	1					33	15	265.0	12
f13.tt	13	1					40	23	9.6	8
f14.tt	14	1					63	27	25.5	12
f15.tt	15	1					117	131	103.7	11
f16.tt	16	1					94	95	23.7	10
f17.tt	17	1					392	257	193.2	11
f18.tt	18	1					224	199	53.6	12
f19.tt	19	1					347	287	6.6	13
f20.tt	20	1					762	575	185.7	15

Self-Organizing Kohonen Maps for FPGA Placement

David C. Blight and Robert D. McLeod

Department of Electrical and Computer Engineering University Of Manitoba
Winnipeg, Manitoba Canada R3T 2N2

Abstract. In this paper we present a novel placement algorithm for
FPGAs. This algorithm is based upon the self-organizing map used in
unsupervised learning algorithms for artificial neural networks performing pattern classification. The self-organizing map is used to map the
connectivity of the design to a two dimensional regular mesh topology.
This is followed by simple compaction to minimize wire costs.

1 Introduction

The popularity of Field Programmable Gate Arrays (FPGA) has been dramatically increasing in recent years. Consisting of a large array of identical programmable Lookup-Tables (LUTs) blocks connected by programmable interconnects, these devices may be programmed to implement both combinational
and sequential circuits. Although each FPGA device contains a large number
of LUTs, it is not always possible to utilize all of the available resources for a
particular design. In order to increase the utilization of available resources it is
important to develop tools which can efficiently synthesize design specifications
into a programmed device. FPGA tools typically divide this process into three
steps. First the design specification is partitioned into simple functions, each of
which is implementable by the LUTs in the device. Next each function is assigned to a specific LUT in the FPGA device. Finally the interconnects between
functions are routed. In this paper we focus on the placement problem of assigning functional tasks to individual LUTs in the FPGA such that constraints
based on area and performance are optimized.

Placement algorithms utilized in CAD tools for FPGAs have traditionally
employed algorithms originally developed for other technologies including Printed
Circuit Boards (PCB) and Integrated Circuit (IC) layouts. Although the placement problem for FPGA is closely related to those of other technologies, the
basic structure of FPGA is different from other technologies. FPGA are implemented as a prefabricated array of programmable logic blocks and interconnects.
The regular structure of such devices make the use of alternative placement algorithms attractive. In this paper we present a placement algorithm developed for
FPGAs based on the unsupervised learning algorithms used in Artificial Neural
Networks (ANN) [2]. The Kohonen self organizing map [1] is used to first map
the connectivity of the design to a two dimensional regular mesh topology, and

then simple one and two dimensional compaction algorithms may be used to produce an area efficient and highly routable mapping.

The remainder of this paper is organized as follows. Section 2 introduces the Kohonen self organizing map. Section three presents placement algorithms for FPGAs utilizing the Kohonen map algorithm. The next section presents some results and comparisons of the new algorithm with the commonly used simulated annealing algorithm [3]. The final section discusses the usefulness of this algorithm as well as future work.

2   Kohonen Self Organizing Map

The Kohonen self organizing map is an unsupervised learning algorithm for ANNs initially developed for applications including speech and pattern recognition. Although FPGA placement problem does not appear to be closely related to the pattern recognition problem, the self organizing map produces a spatially ordered map of its input signals. It is this spatial ordering which we will utilize in the placement algorithm.

The Kohonen map exhibits two essential effects which produce the spatial ordering.

- Spatial concentration of neural activity in a neighborhood which best matches the input. This is a product of the competitive nature of the the learning algorithm.
- Weight adjustment of the best matching neuron and its topological neighborhood.

2.1   Neural Model

The basic structure of the neural network used in Kohonen maps is show in Figure 1. A two dimensional array of neurons are layed out in a rectangular fashion. Each neuron is connected to the input signals of the network. The input signals can be represented as an n-dimensional vector.

$$x = [x_0, x_1, \ldots, x_{n-1}] \in \Re^n \tag{1}$$

Each neuron in the network is connected to each of the input signals, and each connection has a weight associated with it. The weights for each neuron i can also be represented as a vector.

$$m_i = [m_{i0}, m_{i1}, \ldots, m_{in-1}] \in \Re^n \tag{2}$$

2.2   Learning Algorithm

Learning refers to the determination of weight values on connections in a neural network. Although it may be possible to analytically determine the optimal

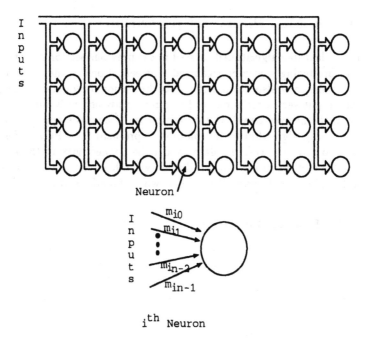

Fig. 1. Kohonen Neural Network

weights for a network, it is usually more convenient to employ learning algorithms which allow networks to determine their weights based on training sets of input/output signals. Unsupervised learning differs from the learning algorithms traditionally employed in ANNs in that no output values are specified in the training set.

The learning of weights proceeds by repeatedly selecting an input vector from the training set (we will discuss in the next section what the training set is), and finding the best match between neuron weight vectors and the input vector (training vector labelled x). The best match refers to the neuron (j) in the network such that :

$$\|x - m_j\| = min\left(\|x - m_i\|\right) \qquad (3)$$

Once the best match is found, the weights of the connections to the neuron and those of its neighbors are adjusted according to the following:

$$m_i(t+1) = m_i(t) + h_{ji}(t)\left[x(t) - m_i(t)\right] \qquad (4)$$

$$h_{ji}(t) = h_0(t)e^{-\|r_i - r_j\|/\sigma(t)^2} \qquad (5)$$

where $\|r_i - r_j\|$ is a measure of the distance between neuron i and j (the best match). $h_0(t)$ and $\sigma(t)$ are function decreasing with time.

3 Kohonen Map Placement Algorithm

The Kohonen learning algorithm discussed in the previous section has the special property of producing spatially organized representations of the input training set. This process may also be thought of as dimensional reduction. The higher dimensional input vector set is reduced to a representation in 2 dimensions (the dimension of the neuron array). In this section we discuss the application of the Kohonen learning algorithm to the placement problem for FPGAs. The basic approach is to have the training data model the connectivity of the circuit to be placed. The corresponding map produced by the Kohonen learning algorithm may then be used to determine the LUT to which functions are to be assigned.

3.1 Circuit Representation

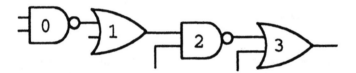

$$X_0 = (1, 1/2, 1/3, 1/4, \ldots)$$
$$X_1 = (1/2, 1, 1/2, 1/3, \ldots)$$
$$X_2 = (1/3, 1/2, 1, 1/2, \ldots)$$
$$X_3 = (1/4, 1/3, 1.2, 1, \ldots)$$

Fig. 2. Circuit Representation

Let N be the number of gates in the circuit to be placed. A gate refers to a function which is to be implemented by a LUT. The circuit C may be represented as a graph G,

$$G = (V, E) \tag{6}$$
$$V = (v_0, v_1, \ldots, v_{n-1}) \tag{7}$$

where V, the vertices of the graph, is the set of gates in the circuit. E is the set of edges in the graph. An edge exists between two vertices if there is a direct connection between the corresponding gates. An example is shown in Figure 2.

The set of training vectors T, is a set of n-dimensional vectors:

$$T = (x_0, x_1, \ldots, x_{n-1}) \tag{8}$$

$$x_i = (x_{i0}, x_{i1}, \ldots, x_{in-1}) \tag{9}$$

such that

$$x_{ii} = 1.0 \tag{10}$$

$$x_{ij} = \frac{1}{d(v_i, v_j) + 1} \quad for\ i \neq j \tag{11}$$

where $d(v_i, v_j)$ is the distance between vertices v_i and v_j.

3.2 Basic Learning Algorithm

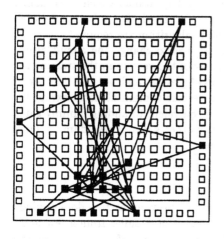

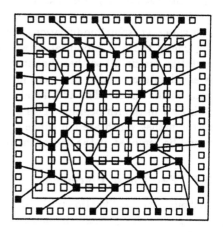

(a)   (b)

Fig. 3. Kohonen Placement

The basic Kohonen learning algorithm described in the previous section is usable with minor modifications. The basic layout of a Xilinx type FPGA [4] is shown in Figure 3. The device contains a square array of Combinational Logic Blocks (CLBs) surrounded by a ring of I/O Blocks (IOBs). Each block in the FPGA device corresponds to a neuron in the Kohonen network.

In order to prevent gates requiring I/O blocks from being assigned to CLBs, and gates implementing CLB function from being assigned to IOBs, the best match criteria is modified to distinguish between IOBs and CLBs. If a training vector corresponds to a gate requiring an IOB, only the IOBs in the device are searched for the best available match. Likewise, only CLBs are searched for the best match for training vectors corresponding to CLBs. The distinction does not affect the selection of neighborhoods around blocks.

An example of a placement of a simple regular circuit structure is shown in Figure 3. The circuit being placed is a simple 5 by 5 mesh of gates, with all edge gates connected to outputs. This example was chosen because it allows the behavior of the algorithm to be clearly seen. Each black square in the figure indicates that a gate has been assigned to that block. This means that the vector associated with one of the gates has that block as its best match. It can be seen that the layout topology matches the original topology.

3.3   Extended Circuit Representation

Although the layout in Figure 3 looks nice, it is not the optimal layout in in terms of any measure of placement quality. It is important to realize that the goal of the Kohonen algorithm is not specifically to optimize the placement according to some cost function, but instead to find a topological map which corresponds to the training set. In order to produce a placement which minimizes the area and wire cost, some modifications to the basic algorithm are required.

One of the first modification we explored was to change the circuit representation for circuit so that the dimensionality of the vectors in the training set is equal to the number of blocks in the FPGA device, and not the number of gates in the circuit. The additional training vectors contain all zeros except for a single 1 in the ith position. This process is equivalent to adding dummy gates to the netlist which are not connected to any other gates. These dummy gates would be removed at the end of the placement.

3.4   Two Phase Placement

Although the modification to the algorithm in the previous section does produce a better placement than the original algorithm, it does suffer from a few practical problems. First of all it dramatically degrades the performance of the the the algorithm by increasing the dimensionality of the problem. Secondly, the basic algorithm does not offer any easy method to introduce external constraints in the placement such as routing priorities.

An alternative approach is to separate the placement procedure into two phases. In the first phase, the basic Kohonen placement algorithm is used. In

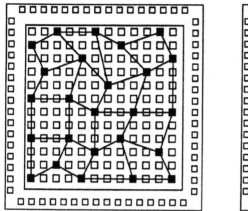 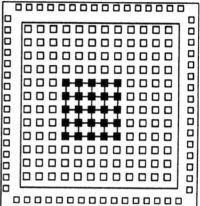

(a)   (b)

Fig. 4. Two Phase Placement

the second phase traditional one or two dimensional compaction algorithms are used. An example using simulated annealing with a temperature of zero for the second phase is shown in Figure 4. Alternative compaction algorithms would be equally suited.

4   Results

In this section we will present some comparisons of the two phase Kohonen placement algorithm. We will compare the algorithm to the simulated annealing algorithm. Our measure of performance will be based on the time to perform placement, and a measure of the total distance of wire required to connect the circuit.

Circuit	Kohonen		Sim Anneal	
	Time	Wire	Time	Wire
Mesh	100	52	250	52
Mesh*	110	68	250	65
ALU	250	120	300	115

Mesh* refers to a mesh with wrap around in the horizontal direction.

5   Conclusions

In this paper we have presented a new placement algorithm for FPGAs based on the Kohonen self organizing map. Initial research in this area has shown that this algorithm when combined with traditional compaction algorithms, may offer improved performance over other commonly used placement algorithms such as simulated annealing. This approach seem particularly well suited for placement problems involving regular structures including meshes. It is not limited to these topologies however.

Future work in this area will focus on extending the algorithm to be more flexible and allow more externally specified constraints. The second phase of the placement algorithm should utilize a more robust compaction algorithm than the simple zero temperature simulated annealing algorithm used. We are also investigating using elastic neural network algorithms in an attempt to have compaction performed simultaneously with the placement.

References

1. Kohonen,T.: The Self Organizing Map. Proceedings of the IEEE. Vol. 78 No. 9 (1990) pp.1464-1480.
2. McClelland,J.L., Rumelhart D.E.: Explorations in Parallel Distributed Processing. MIT Press (1988).
3. Kirkpatrick S., Gelatt C.D., Veechi M.P.: Optimization by simulated annealing. Science 220 (1983) pp. 671-680.
4. Xilinx, The Programmable gate Array Data Book, Xilinx 1991.

High Level Synthesis in an FPGA-Based Computer Aided Prototyping Environment*

P. Poechmueller, H.-J. Herpel, M. Glesner, F. Longsen+

Darmstadt University of Technology
Institute of Microelectronic Systems
Karlstrasse 15, D-6100 Darmstadt

+ Shanghai University of Science and Technology
Physics Department

Abstract. This paper presents a design methodology to support the design of embedded information processing units in mechatronic systems during early design phases. System partitioning into a set of software and hardware modules is done at system description level. User guided and automated synthesis tools generate a fully functional prototype that can be connected to the mechanical subsystem to estimate system performance. The spectrum of realizations ranges from single task software implementations on a single standard processor to application specific integrated processors in a heterogeneous multi-processor environment. In this paper emphasis is put on high level synthesis aspects for the ASIP emulation part of the whole system.

1 Introduction

The overall system performance of machine tools, vehicles and aircrafts can be improved significantly applying embedded information technology and, in particular, artificial intelligence to it. Mechanical systems with embedded information processing elements are usually referred to as 'mechatronic systems'.

Growing complexity of control applications for mechanical systems (e.g. combustion engines, shock absorbers, etc.) requires system realizations which have to execute complex mathematical procedures under hard real time constraints. Verification of embedded information processing algorithms for mechatronic systems in early design phases is essential to avoid time and cost intensive re-designs.

There are many different ways of verifying real time systems during early design phases [1]:

- System simulation

- Breadboarding with standard MSI components

---

*This research was sponsored by the DFG through SFB 241 'IMES'and ESPRIT BRA 3281

- Silicon prototyping

- Computer Aided Prototyping (CAP)

System simulation is mandatory prior to any design activity. Currently several tools are available to support system modelling and analysis of complex mechatronic systems, e.g. MatrixX/SystemBuild, Model-C, XANALOG, StateMate just to name a few of them. All these tools provide a block oriented graphical design entry to describe both, the mechanical system and the embedded information processing unit. Besides pure software simulation real time hardware-in-the-loop simulation is also supported through hardware accelerators with digital and analog I/O capabilities. Final implementation of the graphical model is supported through automatic code generation modules for C, FORTRAN or VHDL. Adaptation of the generated code to the user's hardware environment can be done easily.

Manual Breadboarding with standard MSI chips is not practicable for designs of more than a few thousands gates. Small ASICs of less than 5,000 gates can be prototyped with a few FPGA parts. However, when the design is spread over more than a few FPGAs, partitioning becomes the major problem. There are too few pins on most FPGAs for the number of raw gates inside, so when the design is partitioned accross multiple FPGAs the utilization per FPGA decreases drastically.

Silicon prototyping with short turn arouned times at reasonable costs seems to be attractive for application specific integrated processors. Although it may be possible in the near future to achieve these goals, this approach suffers an obvious drawback: it is extremely difficult to debug silicon in an operating environment. Changes in the design require a complete new fabrication run, which takes time and is cost intensive.

Computer Aided Prototyping combines some of the techniques mentioned above and uses modern synthesis tools to generate a fully functional prototype capable of performing hardware-in-the-loop simulations. The spectrum of realisations ranges from single task software implementations on a single standard processor to Application Specific Integrated Processor (ASIP) based solutions.

2 Overview on the CAP Environment

Our CAP environment consists of a powerful heterogeneous multi-processor system and a set of high level and structural synthesis tools (see fig. 1).

The user enters system description as communicating arithmetic state machines (ASM) either in graphical or ASCII format. A compiler checks the input file for consistency and completeness and generates code for each function specified in the system description (Pascal, assembler, HardwareC). These source files are the starting point for mapping onto different processing nodes (e.g. direct compilatin of Pascal to assembler and consecutive execution on DSP board).

The complete prototype hardware realization provides the capability for real time hardware-in-the-loop simulations and consists of 6 processing nodes embedded in a network of 4 busses splitted into 6 bus segments. Each of the processing nodes can be implemented differently, e.g. a complete system can be realized through 3 DEU-nodes, 1 DSP-node and 2 Interface nodes (one interface node to the process and one to the

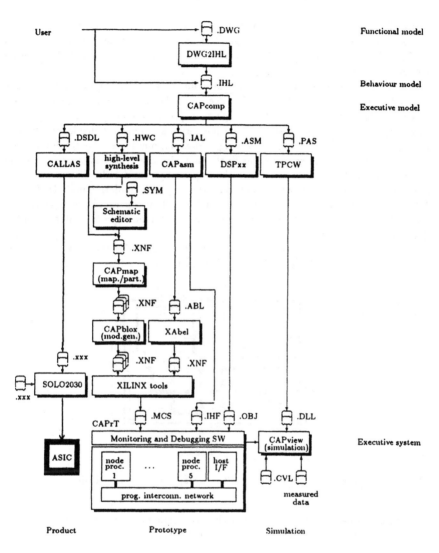

Figure 1: System Modelling and Simulation Environment

host). One of the single DEU-nodes already implements a complete datapath! Two operands can be written into a processing node at the same time. They either can be stored in a dual ported RAM, a FIFO or a mailbox. Operations on these data start when all data are available (data flow oriented computing). A complete system is restricted to max. 6 processing nodes due to interconnection and bus problems.

All processing nodes principally have the same architecture: a motherboard provides program memory (8kx64), data buffers (FIFO, mailbox, DualPort RAM and semaphore logic), download and debugging circuitry for the plugged in data processing unit. The data processing unit (DPU) can be either processor based (e.g.

TMS320C30, FuzzyProcessor FC110) or it can be a datapath emulation unit (DEU). The DEU consists of 4 programmable gate arrays (Xilinx XC4005 series) which can be configured as application specific datapath. The controller is formed by the program memory and another LCA (see fig. 2).

High level synthesis tools [3], [2] are available to generate the datapath and corresponding controller for DEU-boards since a high level synthesis tool supports prototype implementation of ASIPs through field programmable gate arrays (FPGA) including all high level synthesis steps like memory management, behavioural transformations, algorithmic retiming, scheduling/allocation, binding etc. A top-level Xilinx netlist format (XNF) is generated as final result of those synthesis steps. A module generator reads this netlist and fills in the implementation details. This netlist must next be partitioned in a way that it fits onto the datapath emulation unit (DEU) with 4 Xilinx XC4005 FPGAs. The paper will be focused on this rapid prototyping path from a HardwareC description downto its realization on a datapath emulation unit as described in section 3.

An interactive process visualization unit (PVU) provides the user interface to adjust system parameters and observe system response.

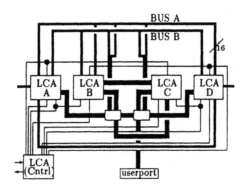

Figure 2: Architecture of the Datapath Emulation Unit

Sensors/actuators interfacing the mechanical system can be connected to the processing nodes through analog and digital I/O channels.

3   High Level Synthesis

Current synthesis algorithms and systems are frequently applied to unrealistic problems as e.g. the well known 5th-order digital elliptic wave filter [4] which is not characteristic for most real world design problems. Corresponding data and control flow graphs are generally small, compact and there are no larger background memory streams to be managed. However, realistic design tasks of mechatronic systems frequently include applications like combustion engine control, clutch control, tire friction control etc. which are much more difficult to synthesize. Therefore, a new synthesis framework devoted to this application domain had been developed and included into the presented Computer Aided Prototyping environment.

This high level synthesis framework is used within the prototyping environment to generate FPGA solutions for single process descriptions in HardwareC. As already mentioned, this HardwareC description can be generated automatically through the prototyping environment after system specification and partitioning. The synthesis process starts from the HardwareC specification of a single process which typically shows the following characteristics:

- data/control flow graphs are large and complex

- there are frequently irregularities like special computations for the first and last values of large vectors

- input data is only one- or two-dimensional (vectors, matrixes, nonterminating signal streams etc.)

- during structural synthesis a large number of operations is mapped onto a small number of hardware modules. This means the hardware sharing factor (HSF) is large (> 100)

- background memory management is crucial due to strict timing constraints and its large size

In the following we will frequently refer to the term HardwareC, however, this specification language is not comparable to Stanford HardwareC. A new HardwareC language had to be developed for this project since Stanford HardwareC proved to be insufficient for the description of the intended mechatronics applications which was mainly due to the fact that there are hard restrictions in variable types (no float and fix types), arrays for complex types are not available in Stanford HardwareC and since the use of at compile time unknown variables is strongly restricted. All these problems had been overcome in the newly developed HardwareC.

Figure 3 gives a rough overview of the synthesis process starting from the HardwareC specification. First synthesis step is a profound data/control flow analysis performed through the HardwareC compiler. The specification analysis results in the transformation of the HardwareC description into a completely analysed data/control flow graph (DCG). The DCG is the prime data structure used during the high level synthesis process an combines both data flow as well as control flow information in a single graph. Figure 4 depicts a small HardwareC fragment and the corresponding DCG-format. The DCG-semantics is based on a single token flow concept and proved to be excellent suited as a base for high level transformations.

The HardwareC compiler already performs graph optimizations as they are known from software compilers. Those optimizations include dead code check, redundancy removal, movement of constant constructs out of loops etc. The next major high level synthesis transformation is background memory management.

3.1 Memory Management

Background memory management is of major importance during high level synthesis of mechatronics applications since algorithmic specifications are frequently abundant in arrays (e.g. matrixes or vectors). A direct realization in hardware would result in prohibitively inefficient solutions.

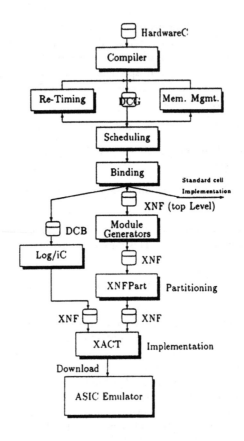

Figure 3: High Level Synthesis for the ASIC Emulator Board

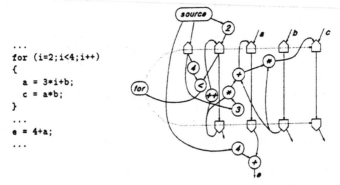

```
...
for (i=2;i<4;i++)
{
  a = 3*i+b;
  c = a*b;
}
...
e = 4+a;
...
```

Figure 4: HardwareC fragment and corresponding DCG

First task to be solved is I/O management. This is the process of assigning n

streams of data to m physical ports. Furthermore, several data streams can be interleaved e.g. 2 vectors $v_1 = (n_1, n_2, ..., n_k)$ und $v_2 = (m_1, m_2, ..., m_l)$ with $l = 2k$ can be read through one physical port in the following sequence:

$$< n_1, m_1, m_2, ..., n_k, m_{l-1}, m_l >$$

This process is mainly done through user interaction since decisions depend on available technology, affordable pin count, system environment etc. However, this process is supported through the I/O manager software which gives information on computations which are delayed through I/O, on critical data streams a.o.

After I/O management all consecutive memory management steps are automatic, however can be influenced through user defined statements in the HardwareC description. The removal of unnecessary arrays is one important processes during memory

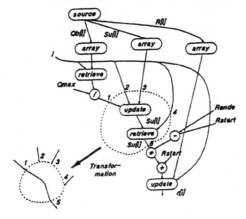

Figure 5: Removal of unnecessary arrays

management replacing many redundant memory write- and read-operations through simple data edges. A very simple example should give a flavor on what kind of transformations are performed at this stage. The following few statements are taken from a complex combustion engine ignition point control algorithm:

```
/* normalization */
Su[i] = Qb[i]/Qmax;
/* air number */
r[i]  = Su[i] * (Rende - Rstart) + Rstart;
```

The value $Qb[i]/Qmax$ is computed and assigned to the array $Su[i]$. The same value $Su[i]$ is directly accessed in the next statement for further processing. Human programmers are frequently using arrays in such a style with the intention to define rather connections instead of storing large data entities. If no other accesses are performed to the value $Su[i]$ in the example above then the complete update/retrieve can be replaced through a data edge as depicted in figure 5. If there are no more update/retrieve operations on an array then even the whole array can be replaced.

Further major background memory management steps are array compression, array merging and selection of actual memory modules. Array compression exploits the fact that during algorithm execution only a small window of the array is actually used (e.g. computing a[i] = a[i-1] + 2*a[i-3] requires a window from a[i..i-3]) while older values are not further used. In that case compression can achieve dramatic size reductions. The same holds for merging different arrays into a single one, however, it has to be taken into account that address generation becomes more difficult requiring additional overhead. Final background memory management step is the selection of real modules like efficient shift registers, RAM-blocks etc. for storing the large data entities.

Table 1 shows the results achieved on a complex combustion engine control algo-

	before	after	reduction
nodes	1047	765	29.9%
edges	1953	1425	27.0%
arrays	37	6	83.8%
updates	48	6	87.5%
retrieves	123	26	78.9%

Table 1: Results on combustion engine algorithm

rithm. After compilation the data/control flow graph comprised approx. 1000 nodes and 2000 edges! In the initial HardwareC specification 37 arrays (vectors of 128 float values each) had been used to define the intended algorithm. Achieved results are equivalent to those of a manual design [5].

3.2   Algorithmic Retiming

Another core high level behavioural transformation is algorithmic retiming which can also be refered to as loop folding [7] [6]. Corresponding methodologies are already long known in the field of optimization of synchronous circuits through shifting of delay operators (registers) [8]. New algorithms had been developed in this project to achieve similar effects on the algorithmic level. Thereby, it is possible to generate pipeline effects if a subgraph executed within one loop iteration is partitioned into several segments executed in different consecutive loop iterations. Corresponding loop transformations are supported through a very fast list scheduler since the purely behavioural graph information is insufficient to steer the retiming process.

Algorithmic retiming can tremendeously improve the quality of synthesized solutions since throughput is increased which is a crucial design parameter in many mechatronics applications. In some of the investigated applications throughput could be increased by a factor of up to 2-3 through retiming!

Besides background memory management and algorithmic retiming other high level transformations like loop unrolling, tree balancing, etc. are implemented which are steered through memory management and retiming results which is a major problem of most other synthesis approaches.

3.3 Structural Synthesis

After high level behavioural transformations the changed DCG will be submitted to structural synthesis. Scheduling and allocation is performed in parallel through the ADaPaS-scheduler which uses advanced genetic algorithms and complex cost functions [9]. Interconnection-, register- and multiplexer-costs are taken into account.

After scheduling/allocation the binding task is performed which assigns variables to registers and register transfers to interconnections resp. busses. A complex clique partitioning/covering framework had been developed for solving associated minimization tasks.

At this design stage a complete netlist is available in XNF-format on the architectural level implementing the original HardwareC specification. The scheduler also produces control information (DCB-format) from which a controler can be produced automatically via commercially available tools like Log/iC. Via technology mappers, module generators and partitioning tools it possible to produce the final XNF-format which can be downloaded onto a datapath emulation unit (DEU).

4 Conclusions

Our design environment supports the verification of embedded information processing algorithms during early design phases through rapid prototyping and true real time hardware-in-the-loop simulations. Switching between different realizations (soft/hardware) of certain functions can be done easily through re-compilation of the system description with another attribute assigned to the function. The user gets a feedback about the system performance through the hardware-in-the-loop simulations. He also gets an idea of the complexity of different realization in terms of RAM requirements for software implementations and in terms of gate complexity for an ASIC based solution.

References

[1] S. Walters, *Computer Aided Prototyping for ASIC-based Systems*, IEEE Design and Test of Computers, June 1991

[2] P. Poechmueller, M. Glesner, *HADES - High Level Architecture Development and Exploration System* First Great Lake Symposium on VLSI, Kalamazoo, Michigan, pp. 342-343, March 1991

[3] F. Catthoor, N. Wehn, P. Poechmueller, et al, *Novel ASIC Architecture and Synthesis Methodologies for Future Multiplexed Datapath Designs*, CompEuro91, Bologna, pp. 506-511, May 1991

[4] L. Stok, R. van den Born, *EASY: Multiprocessor Architecture Optimisation*, Logic and Architecture Synthesis for Silicon Compilers, pp.313-327, North-Holland, 1990

[5] A. Laudenbach, M. Glesner, G. Hohenberg, E. Nitzschke, D. Koehler, *Real Time Heat Release Calculation of Combustion Engines*, Proceedings of 24th ISATA International Symposium on Automotive Technology and Automation, Florence, May 20-24, pp.733-740, 1991

[6] J. Rabaey, M. Potkonjak, *Retiming for Scheduling*, VLSI Signal Processing IV, IEEE Press, pp. 23-32, 199

[7] G. Goosens, J. Vandewalle, H. De Man, *Loop optimization in register transfer scheduling for DSP-systems*, Proceedings of 26th Design Automation Conference, pp. 826-831, Las Vegas, 1989

[8] C.E. Leiserson, F.M. Rose, *Optimizing Synchronous Circuitry by Retiming*, Third Caltech Conference on Very Large Scale Integration, Computer Science Press, 1983

[9] N. Wehn, M. Held, M. Glesner, *A Novel Scheduling/Allocation Approach for Datapath Synthesis based on Genetic Paradigms*, Logic and Architecture Synthesis, pp. 47-56, published by North-Holland, 1991

[10] H.-J. Herpel, N. Wehn, M. Glesner, *RAMSES - A Rapid Prototyping Environment for Embedded Control Applications*, Proceedings Second Int. Workshop on Rapid System Prototyping, Research Triangle Park, June 1991

New Application of FPGAs to Programmable Digital Communication Circuits

Naohisa Ohta, Kazuhisa Yamada, Akihiro Tsutsui and Hiroshi Nakada

NTT Transmission Systems Laboratories
1-2356, Take, Yokosuka-shi 238-03 Japan
E-mail: naohisa@ntttsd.ntt.jp

Abstract This paper proposes a new design method to construct flexible, high performance digital communication systems. The method, called *Amphibious Logic,* combines top-down design with high level synthesis and reconfigurable hardware. The method's capability and problems that had to be solved associated are discussed. Design examples using the high level CAD system called PARTHENON and conventional FPGAs are illustrated. The results show that it is possible to create programmable, high performance digital communication circuits with the proposed method.

1. Introduction

Recent advances in VLSI technology and optical transmission have provided unparalleled opportunities to create broadband digital communication networks. In the future networks, it is likely that new user demands will arise and thus some communication circuits functions may need to be modified. This implies that device programmability and a digital signal transport processing speeds are indispensable for future digital networks so that new services and architectures can be easily implemented. Basing communication circuits on a microprocessor is one way of realizing programmability. However, general microprocessors are not suitable for high speed bit stream manipulation.

In this paper, a new design method to construct flexible, high performance digital communication systems is discussed. The basic idea is that pipeline arrays of FPGAs[1,2] or FPGA-like devices consisting of logic circuits and flip-flops achieve both high speed and programmability. Design examples using the high level CAD system called PARTHENON and conventional FPGAs are also described.

2. Demand for Programmability in Communication Systems

A communication node system can contain numerous kinds of functions which can be divided into layers. In general, the physical layer requires the highest speed but little programmability. Between the application and physical layers are the layers associated with digital signal transport. The logic operations in these layers must achieve:

- mega to giga bit per second serial data transport
- multiplexing, header processing, cross-connect
- other bit-level operations.

All the data handled in these operations are assumed to be synchronous. These operations reflect new types of services and new transport architectures. Considering that services are always changing and that user demand will change some part of the transport architecture, it makes sense to construct communication nodes with programmable digital signal transport circuits.

3. New Design Approach: *Amphibious Logic*

The conventional way of creating programmability is to use processors. We can make a program that represents what we want the communication circuits to do by using a high-level language. The compiler converts the program into a machine language so that the procedure in the program matches the processor's architecture. As the processor has a pre-determined architecture represented by its instruction set, its performance will depend on how well the architecture suits the target operations. Unfortunately, conventional processors are not always good at bit-level logic operations.

The new approach proposed here is a mixture of software and hardware. It can be called *Amphibious Logic* which was named from the analogy of "amphibia." The basic idea is to combine top-down design with logic synthesis and programmable devices. It is expected that, considering recent advances in logic synthesis, we can regard the logic synthesis process as a compiling process in conventional program logic. If appropriate programmable devices are chosen, they would provide high performance as well as programmability. Fig. 1 shows the design procedures of program logic and Amphibious Logic.

There are several problems that must be solved before amphibious-logic-based communication systems can be fully realized. The key subjects are as follows:

(1) Programmable device architecture for digital communication circuits.

(2)High performance logic synthesis and technology mapping tools for the FPGA architecture.

(3)System architectures which incorporate programmable devices as well as processors and hardware.

To resolve these problems, the authors started by studying how feasible amphibious logic was in achieving the target operations in communication circuits. The results are described in the next chapter.

4. Design Example

(1) Behavioral description and logic synthesis

The first step was to verify the current performance of logic synthesis for digital signal transport circuits in communication systems. We used SFL[3] as the behavioral description language and the PARTHENON[3] system as a logic synthesis tool. We chose SFL and the PARTHENON system because they provide a fast logic simulation and synthesis environment for synchronous digital circuits. We made an SFL description of a circuit which had been conventionally designed so that it could be synthesized[4]. The function of the circuit was to process the frame header of the synchronous interface as standardized by CCITT [5]. The circuit performs the typical operations of communication circuits. The functional block diagram is shown in Fig. 2. The complete functions of this circuit can be described in an SFL program of about 900 steps. The synthesized circuit required about 7500 gates while the original manual design used about 6500 gates. It took about 5 minutes for the synthesis on a SPARC station. The workload was about 8 person-days including verification. We have examined various kind of circuits and our observations indicate that PARTHENON designs have less than 50 % more gates than comparable manual designs. This result shows that SFL and PARTHENON are suitable for designing digital communication circuits as expected.

(2) Mapping to Xilinx FPGAs

An automatic design environment was established in order to examine the suitability of the design flow (from logic synthesis to FPGA) and to clarify problems associated with the current FPGA design systems. The PARTHENON system was again used as a top-down CAD tool. Fig. 3. shows the design flow for the Xilinx FPGA [6]. The synthesis data gained from PARTHENON were automatically translated into Xilinx data format. No manual optimization was performed. Various kinds of digital signal transport circuits were examined by comparing manual and automatic design results based on this design flow. The comparison paid attention to the following points.

• Total Number of CLBs (Configurable Logic Blocks: The unit in Xilinx FPGAs)

• Maximum path delay

• Maximum Number of Passed CLBs

The results of PARTHENON logic synthesis depend on maximum fan-in. This is because PARTHENON logic synthesis performs multi-level optimization, which uses the predetermined input number of logic gates, namely the maximum fan-in, with the two-level optimized Boolean equation. We changed the maximum fan-in value from 2 to 5 to examine its influence on mapping.

Table 1 compares some of the design results gained from manual and automatic design flows. Two examples (*fsynch* and *bip8*) are shown. The circuit "*fsynch*", which performs frame synchronization, can be easily described in 120 steps of SFL using a finite state machine representation. The total number of synthesized gates for *fsynch* was 180. As shown in Table 1, the automatic design flow yielded a good design in terms of circuit size. However, for relatively large scale circuits such as fsynch, automatic design may result in a variety of designs. The results vary because the number of maximum fan-in influences the logic synthesis of PARTHENON, and this implies that there is room for improvement by taking the FPGA's architecture into consideration in logic synthesis and optimization.

5. Discussions

In order to realize *Amphibious Logic* in real systems, more research into new FPGA architectures and high performance CAD tools is required. The following requirements should be considered in designing the CAD system.

(1) Logic synthesis should be as fast as possible, equivalent to a compiler, to enjoy the merits of programmability.

(2) Higher speed should come first rather than fewer gates because the target circuits are not for prototyping, but for real operation.

(3) Multi-chip design is indispensable because the target system can be large. Mapping and routing algorithms should support multi-chip design.

(4) The CAD system should incorporate automatic latch insertion to segment the critical timing path. Although the latch insertion shifts the circuit's overall response, it can enhance performance without changing the logic.

6. Conclusions

We have shown a new system design method, called *Amphibious Logic*, which can construct flexible, high performance digital communication systems. Considering the design results described in this paper, it is promising that a top-down design method incorporating logic synthesis for FPGAs can result in high performance communication circuits if an appropriate FPGA architecture is chosen.

References
[1] H.C.Hsieh, K.Duong, J.Y.Ya, R.Kanazawa, L.T.Ngo, L.G.Tinkey, S.Carter, and R.H.Freeman, "A second generation user-programmable gate array," *Proc. IEEE Custom Integrated Circuit Conference*, pp. 515-521, 1987.
[2] A.Haines, "Field programmable gate array with non-volatile configuration," *Microprocessors and Microsystems*, Vol.13, No.5, pp.305-312, 1989.
[3] Y.Nakamura, "An integrated logic design environment based on behavioral description," *IEEE Trans. on Computer-aided Design of Integrated Circuits and Systems*, Vol.CAD-6, No.3, pp.322-336, 1987.
[4] K.Yamada, H.Nakada, A.Tsutsui, and N.Ohta, "A Study on Transport Processing Circuit Design using High Level Design System," *1992 Spring Natl. Conv. Rec. IEICE*, A-102.
[5] CCITT Recommendation G.707,G708,G709: Blue Book, 1989.
[6] Xilinx, Inc., *Programmable Gate Array Data Book*, 1991.

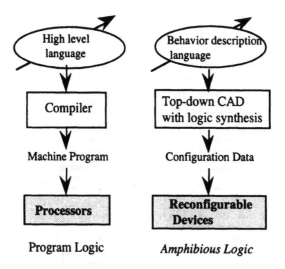

Fig. 1. Design procedures: Program logic and
Amphibious logic

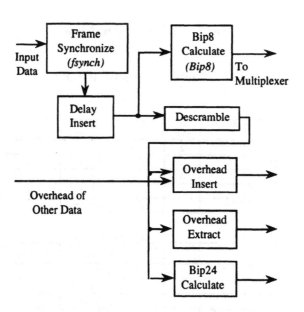

Fig. 2. Block Diagram of Target Function

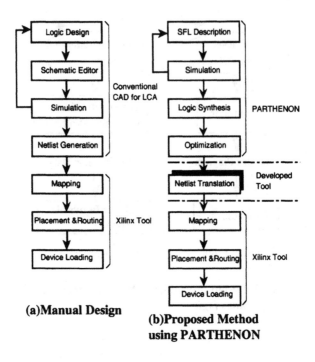

(a)Manual Design

(b)Proposed Method
using PARTHENON

Fig.3   LCA Design Flow

Table 1   Design Results

DEVICE:Xilinx LCA
XC3090PG175-100

(1) fsync

Design Method	Max. Fanin	# of CLBs	Max Path Delay	Max. Passed CLBs
PARTHENON	2	27	71.3ns	3
PARTHENON	3	26	89.9ns	4
PARTHENON	4	28	68.0ns	3
PARTHENON	5	31	65.4ns	3
Manual	-	22	64.1ns	3

2)bip8

Design Method	Max. Fanin	# of CLBs	Max Path Delay	Max. Passed CLBs
PARTHENON	2	8	31.2ns	1
PARTHENON	3	8	31.1ns	1
PARTHENON	4	8	30.9ns	1
PARTHENON	5	8	31.5ns	1
Manual	-	8	30.5ns	1

FPGA Based Logic Synthesis of Squarers Using VHDL

Georg Kempa[1] and Peter Jung[2]

[1] University of Kaiserslautern, Microelectronics Centre (ZMK),
W-6750 Kaiserslautern, Germany
[2] University of Kaiserslautern, Research Group for RF Communciactions,
W-6750 Kaiserslautern, Germany

Abstract. In this paper, the design of VHDL coded squarers by using logic synthesis is considered. The square function is important for the digital processing of signals using e.g. matched filters and Viterbi equalizers in receivers for communication systems. However, many arithmetical functions like the square function are not supported by VHDL. Hence, two major drawbacks arise in the logic synthesis of VHDL code. Firstly, the designers are forced to implement the needed arithmetical functions in VHDL by themselves. Secondly, when implementing arithmetical functions such as the square function in VHDL, special care must by taken in order to circumvent massive hardware overhead of the synthesis results compared with manually designed architectures. In the case of the square function, this massive hardware overhead mainly stems from the fact that the synthesis results of squarers are as hardware expensive as the synthesis results of multipliers. In the course of the present paper, the authors shall demonstrate how this hardware overhead of squarers can be reduced by using a modified square algorithm (MSA) which was developed by the authors. The MSA was derived based on the Dadda algorithm which will be discussed briefly.

1 Introduction

The hardware description language VHDL [1, 2] has become increasingly accepted as a viable tool for the use in the design process because its application by describing the functionality rather than the logic of the design leads to a significant reduction of both the development time and costs. Recently, the possibility of mapping VHDL code on FPGAs by logic synthesis was made available to the designer, thus leading to inexpensive ASIC designs.

With the introduction of digital signal processing in many areas of human life, e.g. communication technology by establishing all-digital radio systems [3, 4], the demand for complex and flexible digital ASICs is considerably increasing. Many important detection algorithms like matched filters [5] and Viterbi equalizers [6, 7] used in digital communication systems incorporate the calculation of the square of some sample value. From an ASIC designer's point of view, the square algorithm is regular and rather simple to implement. Especially, the algorithm introduced by Dadda [8, 9] which will be explained in section 2 leads to favorable squarer implementations.

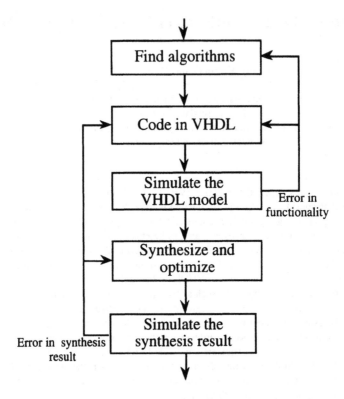

Fig. 1. Design flow using logic synthesis of VHDL code

Nevertheless, the manual design of large squarers is time-consuming. Hence, the automated design using VHDL code and logic synthesis would be desirable provided that the hardware expense of the synthesized hardware does not exceed that of the manually designed hardware. The design flow using logic synthesis as depicted in **Fig. 1** has got several benefits. The most important benefits are the functional verification, in our case by the simulation of bit error rates, designer-friendly modifiability of the VHDL code and its technology independence. Setting out from a given problem, in our case the implementation of a squarer, suitable algorithms have to be found in the first step, cf. Fig. 1. Then these algorithms are encoded in VHDL, see sections 2 and 3 for the square algorithm. Now, the VHDL code can be simulated, revealing the possibility of detecting functional errors. This is a major advantage over the conventional manual design flow. In the conventional manual design flow, the simulation is carried out after the schematic entry is completed. Due to the high computing power required by the simulation of conventional manual designs, a functional verification on the basis of bit error rates is generally impossible. According to Fig. 1, the functionally correct implemented VHDL code is synthesized and optimized. In a second simulation step, errors in the synthesis result are detected. The described design flow which is schematically depicted in Fig. 1 allows for a high reliability of the design. Furthermore, expensive re-designs can generally be avoided.

Unfortunately, the logic synthesis of VHDL code has restrictions primarily stemming from the synthesis tools. Hence, there are restrictions applying to the designer-friendly modifiability of VHDL code that shall be synthesized. Therefore, the designer must assist the synthesis tool by choosing suitably modified algorithms that solve the above-mentioned given problem. In the case of the square function, an advantageous modification of the Dadda algorithm presented in section 2 overcomming some drawbacks of the original Dadda algorithm was found by the authors which allows for both a favorable hardware expense of the synthesized squarers and a short design time. This modified square algorithm (MSA) is presented in section 3.

In section 4, the obtained synthesis results for squarers with different numbers of input bits are given. The comparison of these synthesis results with the manually designed squarers reveals a good hardware efficiency of the synthesized squarers in terms of hardware expense. When mapping the synthesis results onto two different types of FPGAs, namely the XILINX XC3000 series FPGAs which are also called logic cell arrays (LCAs) [10] and the Actel ACT1 series FPGAs [11], considerable differences between the statistics output of the synthesis tool and the final results after the placement and routing are found. These differences shall also be discussed in sectoin 4.

2 Square algorithm after Dadda

The task of a binary squarer is to calculate the square of an integer number a with the binary representation

$$a = \sum_{i=0}^{K-1} a_i \cdot 2^i, \quad a_i \in \{0, 1\}, \quad a, K \in \mathbf{N}. \tag{1}$$

According to (1), a is represented by K bits. The output of the squarer is given by

$$a^2 = \sum_{i=0}^{K-1} a_i \cdot 2^{2i} + \sum_{i=1}^{K-1} \sum_{j=0}^{i-1} a_i a_j \cdot 2^{i+j+1}. \tag{2}$$

The products $a_i a_j$ denote that the logical AND operation is applied to a_i and a_j. The first step according to Dadda is the collecting of all the coefficients of the factor 2^i with i being constant. With $[\cdot]_g$ denoting the integer part of $\cdot$, and with the Kronecker symbol defined by

$$\delta_{i,j} \overset{\text{def}}{=} \begin{cases} 1, i = j, \\ 0, \text{else}, \end{cases} \tag{3}$$

(2) can be displayed as

$$a^2 = \sum_{m=1}^{2K-3} 2^{m+1} \cdot \left\{ \sum_{j=m-K+1}^{\left[\frac{m-1}{2}\right]_g} a_j a_{m-j} + a_{\left[\frac{m+1}{2}\right]_g} \cdot \delta_{\left[\frac{m+1}{2}\right]_g, \frac{m+1}{2}} \right\} + a_0, \quad a_i \equiv 0 \; \forall \; i < 0. \tag{4}$$

In general, the summation inside the brackets in (4) leads to an overflow. This circumstance must be considered in the implementation process. A simple illustration of the squarer structure is depicted in **Fig. 2** for the case of $K = 5$. The coefficients of 2^{m+1} given in (4) refer to the elements of the displayed calculation matrix shown in Fig. 2.

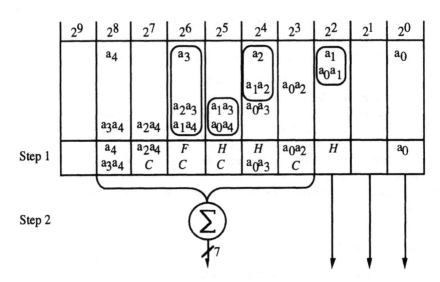

Fig. 2. Conventional implementation of a squarer after Dadda with $K = 5$. The abbreviations H and F denote the results of (2,2)-counters (half adders) and (3,2)-counters (full adders), respectively, whereas C indicates generated carry bits. The output a^2 of the squarer consists of 10 bits. It is determined in two consecutive calculation steps.

The conventional approach is the columnwise adding of the matrix elements by the application of (2,2)-counters (half adders) and (3,2)-counters (full adders) [12]. Therefore, a partitioning of the matrix elements in groups of two or three is necessary. In Fig. 2, the grouping of the appropriate matrix elements is represented by ovals. Usually, this procedure has to be continued in several calculation steps, depending on the number of matrix elements per matrix column stemming from the complexity of the desired squarer. In Fig. 2, only one calculation step referred to as "step 1" applying full and have adders is necessary. In the final calculation step, the result a^2 must be calculated by a rather large binary adder. In the example of Fig. 2, a six bit adder is used in this final calculation step referred to as "step 2". Unfortunately, there does not exist a simple solution of the partitioning problem for large K. Hence, the conventional manual design of a squarer based on the Dadda algorithm is rather elaborate.

For two reasons, the Dadda algorithm presented in this section is not well suited for the logic synthesis of squarers. As mentioned above, depending on the number of input bits K, a different number of calculation steps is needed. If the number of matrix elements in a matrix column, see Fig. 2, is greater than 3, these matrix elements cannot be combined in a single full adder. In this case, at least a second calculation step applying full and half adders is necessary before the final calculation step can be carried out. The variance of the number of calculation steps depending on K is termed structure variance [13]. Therefore, a flexible VHDL coded squarer model dealing with various numbers K of input bits does not exist. This calls the benefits of a design flow using logic synthesis of squarers using the Dadda algorithm into question because the designer-friendly modifiability of the VHDL coded squarer model cannot be guaranteed. The second reason is the large hardware expense of the final adder. Therefore, a suitably modified square algorithm circumventing the considered drawbacks of the Dadda algorithm leads to synthesis results with a better hardware efficiency in a shorter time. Such a modified square algorithm shall be presented in the following section.

3 Modified square algorithm (MSA)

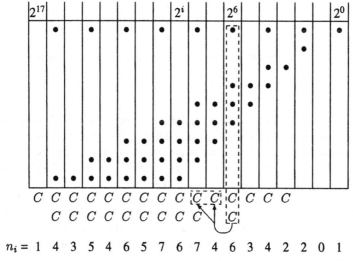

Fig. 3. Simplified representation of the calculation matrix, $K = 9$, $i = 0, 1...17$

In **Fig. 3**, a simplified representation of the calculation matrix for $K = 9$ is shown. Each dot (•) is associated with a binary matrix element a_i or $a_i a_j$, see Fig. 2. The calculation matrix consist of 18 columns, numbered $i = 0$ until $i = 17$. The numbers n_i of matrix elements per column including the carry bits (C) are also shown in Fig. 3. In contrast to the Dadda algorithm, all matrix elements of a column are now added in a single step not only using simple full and half adders. Therefore, this modified square algorithm (MSA) does not exhibit structure variance. This measure allows for a rather straight-forward implementation of (4). For an adder used to add the matrix elements of column i the term column adder (CA_i) is proposed. This kind of summation is schematically depicted in Fig. 3 for column $i = 6$. In this case, a column adder for $n_6 = 6$ input bits is applied.

Column adders for more than 3 input bits generate more than one carry bit besides the resulting bit which is identical with the least significant output bit. In what follows, the number of generated carry bits per column adder CA_i is denoted by $n_{C,i}$. Therefore, the total number of output bits of CA_i is given by $(n_{C,i} + 1)$. The generated carry bits of CA_i must be considered by the column adders CA_{i+1}, $CA_{i+2}...CA_{i+n_{C,i}}$. For instance, $n_{C,i} = 2$ carry bits are generated by CA_6 schematically shown in Fig. 3 by a dashed box. These $n_{C,i} = 2$ carry bits are input bits of CA_7 and CA_8. In **Tab. 1**, the number $n_{C,i}$ of carry bits generated by column adders with n_i input bits is given.

n_i	$n_{C,i}$
2, 3	1
4...7	2
8...15	3
16...31	4
⋮	⋮
$2^\nu...(2^{\nu+1} - 1)$	ν

Table 1. Number $n_{C,i}$ of carry bits generated by column adders with n_i input bits

As discussed above, the MSA does not exhibit structure variance. However, the second problem mentioned in section 2, namely the large hardware expense of complex adders — represented by the final adder in the case of the Dadda algorithm — is not solved. Due to the large variance of n_i, see Fig. 3, both complex and simple column adders are needed. In order to discard the complex column adders, the reduction of the variance of n_i is desirable. Furthermore, in order to guarantee a low hardware expense of the synthesized squarer, the mean value of n_i must be lowered because by this measure the complexities of the necessary column adders can be further reduced. The reduction of both the variance and the mean value of n_i can be achieved by suitable transformations of the calculation matrix.

Fig. 4 schematically shows two possible transformations of the calculation matrix which were used by the authors. In **Fig. 4(a)**, all the matrix elements including the possibly generated carry bits are depicted as dots ($\bullet$). The depicted situation shows two neighbouring matrix columns with $n_i = 8$ and $n_{i-1} = 4$. One element of column i is then transferred to column $(i - 1)$, denoted by the two arrows and the two circles ($\circ$) representing twice the very element that shall be transferred, according to

$$..., n_{i+n_{C,i}}, ..., n_i, n_{i-1}, ... \quad \Longrightarrow \quad ..., (n_{i+n_{C,i}} - 1), ..., (n_i - 1), (n_{i-1} + 2), ... \quad (5)$$

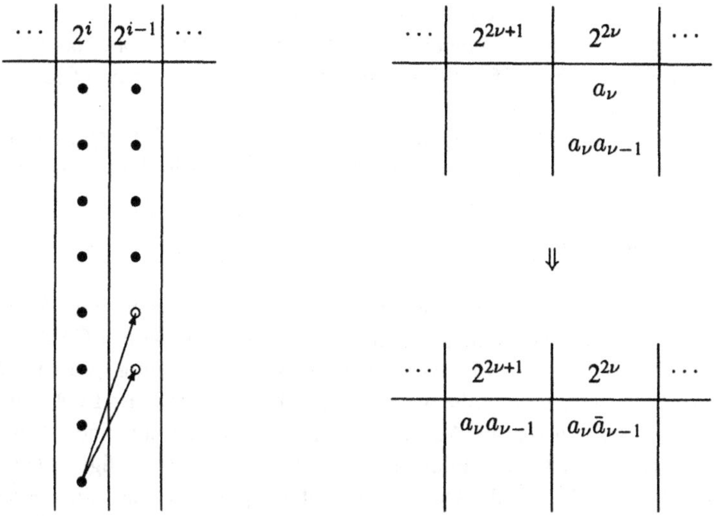

Fig. 4. Transformations of the calculation matrix

Although the number of elements in columns i and $(i-1)$ is increased by one, the number of carry bits is reduced by one according to Tab. 1. Therefore, the variance of n_i can be reduced whereas the mean value of n_i is unchanged by this transformation of the calculation matrix.

In **Fig. 4(b)**, another transformation which is based on the identity

$$
\begin{aligned}
(a_\nu + a_\nu a_{\nu-1}) \cdot 2^{2\nu} &= (a_\nu + a_\nu \wedge a_{\nu-1}) \cdot 2^{2\nu} \\
&= (a_\nu \wedge a_\nu a_{\nu-1}) \cdot 2^{2\nu+1} + (a_\nu \oplus a_\nu a_{\nu-1}) \cdot 2^{2\nu} \\
&= a_\nu \wedge \left(a_{\nu-1} \cdot 2^{2\nu+1} + (1 \oplus a_\nu a_{\nu-1}) \cdot 2^{2\nu}\right) \\
&= (a_\nu \wedge a_{\nu-1}) \cdot 2^{2\nu+1} + (a_\nu \wedge \bar{a}_{\nu-1}) \cdot 2^{2\nu} \\
&= (a_\nu a_{\nu-1}) \cdot 2^{2\nu+1} + (a_\nu \bar{a}_{\nu-1}) \cdot 2^{2\nu}
\end{aligned}
\tag{6}
$$

with $\wedge$, $\oplus$ and $\bar{}$ denoting the AND, EXOR and NOT operations is used. The change of n_i according to the transformation depicted in Fig. 4(b) is given by

$$
\ldots, n_{(2\nu+1)}, n_{2\nu}, \ldots \implies \ldots, (n_{(2\nu+1)} + 1), (n_{2\nu} - 1), \ldots
\tag{7}
$$

in the case of an unchanged number $n_{C,2\nu}$ of carry bits generated by $CA_{2\nu}$ or given by

$$
\ldots, n_{(2\nu+n_{C,2\nu})}, \ldots, n_{(2\nu+1)}, n_{2\nu}, \ldots \implies \ldots, (n_{(2\nu+n_{C,2\nu})} - 1), \ldots, (n_{(2\nu+1)} + 1), (n_{2\nu} - 1), \ldots
\tag{8}
$$

in the case of a decremented by one number $n_{C,2\nu}$ of carry bits. In the second case, the total number of matrix elements is reduced by one. Therefore, this transformation allows for the reduction of both the variance and the mean value of n_i.

Using the described MSA together with the two discussed transformations leads to favorable synthesis results. This issue shall be discussed in the following section.

4 Synthesis Results

Using the MSA described in section 3, VHDL code for squarers with $K = 2, 3, ..., 10$ input bits was prepared and mapped onto both the XC3000 series LCAs from XILINX [10] and the ACT1 series FPGAs from Actel using the Mentor Graphics AutoLogic synthesis tool. In the case of the mapping onto the XC3000 series LCAs, the results of the logic synthesis using the AutoLogic synthesis tool are contained in two separate statistics files, the first one holding the logical function of the configurable logic blocks (CLBs) of the LCAs and the second one containing a list of the corresponding gates. From this second statistics file, the hardware expense in equivalent NAND2 gates can be calculated. In the case of the mapping onto the ACT1 series FPGAs, there is only one statistics file generated by AutoLogic. This particular statistics file contains a list of the gates form the Actel ACT1 library which were used for the implementation of the considered squarer. The contents of this statistics file are needed for the calculation of the hardware expense in equivalent NAND2 gates.

In order to evaluate the synthesis results of squarers using the MSA introduced in section 3, the hardware expense of these synthesis results is compared with the hardware expense of conventional manual designs of squarers using the Dadda algorithm described in section 2. **Fig. 5** shows the obtained hardware expenses of squarers with $K = 2, 3, ..., 10$ for both the design flow using logic synthesis and the conventional manual design flow. One can conclude that the hardware expense of the investigated synthesized squarers using the MSA is of the same order of magnitude as the hardware expense of the manually designed units using the Dadda algorithm. For $K \leq 6$, the hardware expense of the synthesized squarers using the MSA is even below that of the manually designed squarers. The reason for this advantage in hardware expense is the independence of full and half adders of the logic synthesis due to the minimization of boolean equations by the synthesis tool.

According to Fig. 5, the hardware expense for the synthesized squarers is increasing more rapidly than that of the manually designed squarers because the effort of the optimization procedure carried out by the synthesis tool, see Fig. 1, is exponentially increasing with incerasing design complexity, i.e. increasing K in our case. Since the computation resources are limited, only a restricted set of possible designs can be evaluated by the synthesis tool, thus leading to suboptimum implementations in terms of hardware expense. Nevertheless, in contrast to the elaborate conventional manual design, the elapsed design time for the coding of the design flow using logic synthesis, see Fig. 1, took only about a week. Therefore, the design flow using logic synthesis is advantageous for designs with moderate complexity.

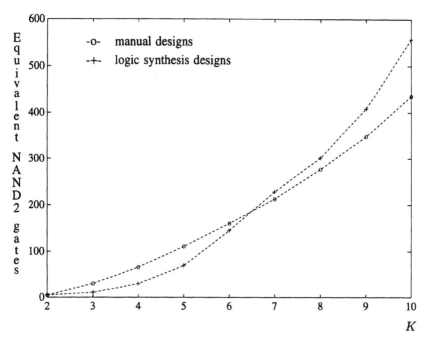

Fig. 5. Hardware expense of both manually designed squarers using the Dadda algorithm and synthesized squarers using the MSA

Fig. 6 shows the number of the used CLBs of the XC3000 series LCAs for the synthesized squarers with $K = 2, 3, ..., 10$. The lower curve (-+-) shows the equivalent NAND2 gates for the synthesized squarers depicted in Fig. 5 converted into CLBs by the division by 9.4. This value is the average based on the results of the MCNC benchmark implementation after [14] using the standard XILINX software xnfmap [10]. The upper curve (-o-) represents the statistics output of the AutoLogic synthesis tool. The middle curve (-*-) gives the used CLBs after the implementation with xnfmap and placement and routing with apr [10]. The three curves of Fig. 6 show large differences. Since the LCA architectures differ from the architectures of conventional gate arrays and standard cells, the difference between the lower curve (-+-) and the middle curve (-*-) is not surprising. However, since the XILINX database for AutoLogic does not fully match the requirements of the logic synthesis with AutoLogic, the difference between the upper curve (-o-) and the middle curve (-*-) of Fig. 6 arises. Therefore, a prediction of the final hardware expense using the AutoLogic statistics file is rather futile. Hence, the benefit of the AutoLogic logic synthesis of LCA implementations is called into question.

In **Fig. 7**, the number of the used logic modules (LMs) of the ACT1 series FPGAs for the synthesized squarers with $K = 2, 3, ..., 10$ is depicted. The lower curve (-+-) shows the equivalent NAND2 gates for the synthesized squarers depicted in Fig. 5 converted into LMs by the division by 5.4 [14]. In contrast to the situation shown in

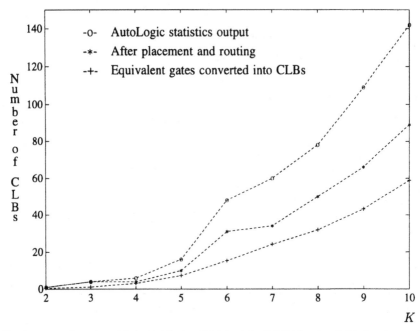

Fig. 6. Usage of configurable logic blocks (CLBs) of the XC3000 series LCAs by the synthesized squarers using the MSA

Fig. 6, the AutoLogic statistics output (-o-) are in nearly perfect accordance with the
used LMs after implementation, placement and routing with als [11] (-*-). Therefore,
the AutoLogic logic synthesis of Actel FPGA implementations is recommendable.

5 Conclusions

In this paper, the design flow using logic synthesis of VHDL-coded squarers is demonstrated. The square function is important for the digital processing of signals using
e.g. matched filters and Viterbi equalizers in receivers for communication systems.
However, many arithmetical functions like the square function are not supported by
VHDL. Therefore, the designers are forced to implement the needed arithmetical
functions in VHDL by themselves. A rather efficient implementation of the square
function is represented by the Dadda algorithm [8, 9] which is briefly discussed in
section 2 of the present paper. However, due to the inherent structure variance and
the large hardware expense of the final adder, the Dadda algorithm is not well suited
for the logic synthesis of squarers. In section 3, the authors present a novel modified
square algorithm (MSA) which circumvents both the structure variance and the large
hardware expense of the final adder.

Based on the MSA introduced in section 3, squarers were synthesized and the synthesis results were mapped onto two different types of field programmable gate ar-

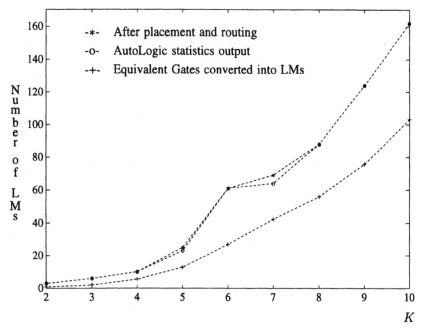

Fig. 7. Usage of logic modules (LMs) of the ACT1 series FPGAs by the synthesized squarers using the MSA

rays (FPGAs), namely the XILINX XC3000 series LCAs and the Actel ACT1 series FPGAs which are used due to their rather low price and their relatively high complexity. The comparison of the hardware expense of the synthesis results with that of conventional manual designs results in comparable amounts of equivalent NAND2 gates.

However, due to discrepancies between the AutoLogic statistics output of the used CLBs and the actually used CLBs after implementation, placement and routing with the XILINX tools xnfmap and apr call the benefit of the AutoLogic logic synthesis of LCA implementations into question. In contrast to the LCA implementations, the used LMs according to the AutoLogic statistics output are in nearly perfect accordance with the used LMs after implementation, placement and routing with the Actel tool als. Therefore, the use of Actel FPGAs together with the AutoLogic logic synthesis of VHDL code is recommendable.

Acknowledgement

The authors wish to thank Dipl.-Ing. (FH) Hans-Peter Goldhammer and Mr. Meinrad Fiedler for their valuable support during the preparation of the data presented in section 4.

References

[1] I.E.E.E. Standard VHDL Language Reference Manual, *I.E.E.E. Std. 1076-1987*, The Institute of Electrical and Electronics Engineers, Inc., 1988

[2] Armstrong, J. R.: *Chip-Level Modeling With VHDL*, Prentice-Hall, Englewood Cliffs, 1989

[3] CEPT/GSM Recommendations, Series 05, 1988

[4] Special Issue on Digital Cellular Technologies, I.E.E.E. Trans. Veh. Technol. **40** (1991)

[5] Proakis, J. G.: *Digital Communications*, 2nd edition, McGraw-Hill, New York, 1989.

[6] Forney, G. D.: Maximum-Likelihood Sequence Estimation of Digital Sequences in the Presence of Intersymbol Interference. I.E.E.E. Trans. Inf. Theory **18** (1972) 363–378

[7] Jung, P., Baier, P. W.: VLSI Implementation of Soft Output Viterbi Equalizers for Mobile Radio Applications. Proc. I.E.E.E. Veh. Technol. Conf. VTC-92, Denver, CO, (1992) 577–585

[8] Dadda, L.: Some Schemes for Parallel Multipliers. Alta Frequenza **34** (1965) 349–356

[9] Dadda, L., Ferrari, D.: Digital Multipliers: A Unified Approach. Alta Frequenza **37** (1968) 1079–1086

[10] The Programmable Gate Array Data Book, XILINX, 1991.

[11] ACT Family Field Programmable Gate Array Data Book, Actel, 1990.

[12] Spaniol, O.: *Arithmetik in Rechenanlagen*, Teubner, Stuttgart, 1976.

[13] Wendt, S.: *Nichtphysikalische Grundlagen der Informationstechnik.* Second edition, Springer, Berlin, 1991.

[14] Weinmann, U., Kunzmann, A., Strohmeier, U.: Evaluation of FPGA Architectures. in Moore, W., Luk, W. (Eds.): *FPGAs*. Abingdon: EE&CS Books, (1991) pp. 147-156

Optimized Fuzzy Controller Architecture for Field Programmable Gate Arrays

Hartmut Surmann, Ansgar Ungering and Karl Goser

University of Dortmund, Faculty of Electrical Engineering, *
44221 Dortmund, Germany,
E-mail: surmann@luzi.e-technik.uni-dortmund.de

Abstract. This paper describes an optimized fuzzy controller (FC) architecture and its realization with field programmable gate arrays (FPGAs). In consideration of data dependencies and minor user restrictions within the definition of fuzzy rules (FRs), it is possible to develop a high speed FPGA architecture. A prototype of the FC operates at 5MHz and needs $50\mu s$ operation time (8 bit resolution) independent of the number of inputs/outputs with 256 fuzzy rules. A pipeline architecture is used to achieve a high processing speed.

1 Introduction

Since Mamdani's work [1] on fuzzy control, which was motivated by Zadeh's approach to inexact reasoning, a lot of work has been reported in this research field so far. The basic idea of this approach was to incorporate the control know-how of a skilled human operator by fuzzy sets and fuzzy rules. The FRs are combined by the fuzzy implication and the compositional rule of inference.

The difficulty of the control know-how is due to their non-linear, time varying behaviour and the poor quality of the available measurements. Fuzzy logic replaces "true" and "false" with continuous membership values ranging from zero to one, which mirror natural language concepts. This allows to process linguistic concepts (adjectives, adverbs) like "small", "big", "near", or "approximately" in the control system. The main advance is to control processes which are too complex to be mathematically modelled in real time.

In the first fuzzy applications FC's are not optimized implemented on standard microprocessors which are flexible, but these first implementations prevent high speed computing. The first approach to get higher performance was to build special microprocessors (RISCs, like FC-110, 80C166) with a special fuzzy instruction set. A much cheaper and faster approach in terms of processing and implementation time is to consider data dependencies and to implement an optimized controller algorithm on a fast 32-bit RISC or standard microprocessor [2]. In addition to the advantages of a FC a very high processing speed and low hardware costs are necessary to achieve a wide acceptance.

---

* This work is supported by the VW-Stiftung, Hannover

In this paper we present an architecture of a general purpose FC and a prototype realization with 256 fuzzy rules, 4 inputs variables and 1 output variable (8 bit resolution). The regular and modular structure of the controller also motivates a VLSI implementation [3, 4, 5, 6, 7, 8].

2 Basic terms of Fuzzy Controller

The FC algorithm is based on the generalized modus ponens inference rule [3]:

Premise:	A is true
Implication:	If A then B
Conclusion:	B is true

The "crisp" propositions A and B are replaced by fuzzy functions. Fuzzy functions characterise and define fuzzy sets through $\mu_i : U \rightarrow [0, 1]$ with $e \mapsto \mu_i(e)$, so $i = \{(e, \mu_i(e)) | e \in U, \mu_i(e)\}$. For fuzzy sets Zadeh [9] defines for $e \in U$ three important fuzzy operations:

Intersection $C = A \cap B$ Union $C = A \cup B$ Complement $\bar{A}$

$\mu_C(e) = \min(\mu_A(e), \mu_B(e))$ $\mu_C(e) = \max(\mu_A(e), \mu_B(e))$ $\mu_{\neg A}(e) = 1 - \mu_{\bar{A}}(e)$

A, B and C are fuzzy sets and U is the universe of discourse for e. These fundamental operations together with the set [0,1] forms a fuzzy algebra, so that any logic function can be build. Instead of $\mu_A(e)$ we only write A to denote the fuzzy set A. The Boolean algebra is a subset of the fuzzy algebra and can be implemented by replacing continuous functions with unit pulses. In difference to a conventional knowledge based system, the premise of the rule is a value in [0,1] instead of {0,1}. The example [10] in Fig. 1 introduces the basic FC algorithm. It shows three simple rules for charging batteries with two inputs dU (gradient of voltage), T (temperature) and one output I (current).

R1: IF dU is negativ and T is normal THEN I is low
R2: IF dU is positiv and T is high THEN I is low
R3: IF dU is positiv and T is normal THEN I is high

The fuzzified inputs dU and T are simultaneously switched to all the rules to be compared with the stored premises (IF parts). Now the truth values α_i^{dU}, α_i^T for every subpremise are calculated by:

$$\alpha_i^{dU} = \mu_{A_i}(dU) \quad , for \ A_i \in \{negativ, positiv\}$$
$$\alpha_i^T = \mu_{B_i}(T) \quad , for \ B_i \in \{normal, high\}, \quad for \ i = 1 \ldots 3 \tag{1}$$

$\alpha_1^{dU} = 0.0$ indicates that the input completely mismatches with the stored subpremise which leads to a complete noncontribution of rule 1 to the output. $\alpha_2^{dU} = 0.85$ and $\alpha_2^T = 0.65$ in rule 2 generates a rule matching or truth value of $\omega_i = 0.65$, because the fuzzy logic conjunction "and" is interpreted as the minimum of α_i^{dU} and α_i^T ($\omega_i = \min(\alpha_i^{du}, \alpha_i^T)$). The conclusion of each rule is

$$I_i' = \{\min(\omega_i, x) \mid x \in I_i\}, \ for \ i = 1 \ldots 3 \tag{2}$$

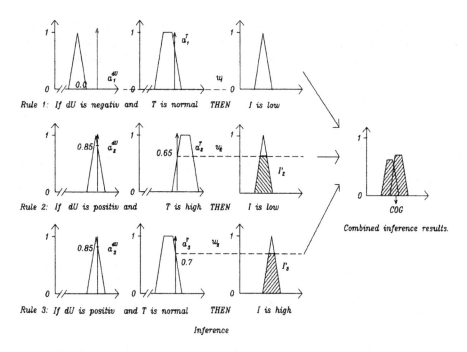

Fig. 1. Fuzzy controller algorithm

and represents the THEN part of each rule. The fuzzy result function I' is the unification of all subresults I'_i and is calculated by $I' = \bigcup I'_i$, for $i = 1 \ldots 3$. In most applications the output values are "crisp" numbers (unit pulses), which are accomplished by calculating the center of gravity (COG) of the resulting fuzzy function I':

$$COG_I = \frac{\int e \times I'(e) de}{\int I'(e) de} \tag{3}$$

The described FC with binary input and output values is called BIOFAMs (Binary Input-Output Fuzzy Associative Memory) [11] or MIN-MAX algorithm [1] with the COG used as defuzzification method.

3 The architecture

A typical FC consists of four different units:

- the fuzzy rule base,   - the fuzzifier unit,
- the inference unit,   - the defuzzification unit.

Only parallel architectures can achieve the realtime requirements. First digital implementations [3] have a high flexibility but need a large amount of chip area because they store the same membership functions several times. After an analyse of the applications, new architectures [5, 8] restrict the degree of overlab and the number of the membership functions. Figure 2 shows the basic structure of the architecture. The FC has a three stage pipeline and two different

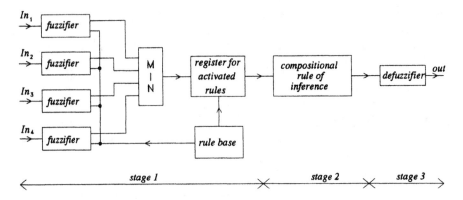

Fig. 2. Pipeline architecture of the fuzzy controller

clock cycles (internal main clock Φ, i.e. 5 MHz and external $\phi = \Phi/256$) which are not shown in Fig. 2. The main clock Φ controls the data flow between the pipeline blocks. The first block process one fuzzy rule each clock cycle ϕ, so that this architecture can handle 256 rules. The results of the premise calculation are stored in a special register block and will be transferred in the next block with the next main clock cycle. There, the output membership functions are limited and the composition is evaluated. The last block (defuzzifier) calculates the "crisp" output value.

3.1 The rule base

The transfer function of a FC is supplied by the definitions in the fuzzy rule base. This rule base connects the membership functions with the "crisp" input values and defines which membership function is activated. The storage capacity S_R for n input and m output variables with k membership functions per input/output variable is $S_R = [\mathrm{ld}(k)](m + n)$ (Tab. 1).

Table 1. Binary code (b) of a fuzzy rule base (a), n=3, m=1, k=8.

Rule 1: if A' is A3 and B' is B4 and C' is C7 then X4   011 100 111 100
Rule 2: if A' is A2 and B' is B4 and C' is C3 then X6   010 100 011 110

    . . .   .

    (a)   (b)

3.2 The Fuzzifier unit

In the fuzzifier unit the inputs are compared with the stored premises (IF parts) and the truth values are calculated for every subpremise. A simple and fast method to store the premises (membership functions) is to use a RAM. Therefore, the input value is the address of the RAM blocks. Eichfeld et al. [5] described an optimized memory organisation in which the membership functions get a binary number and the overlapped membership functions are stored in different RAMs (Fig. 3). Three different memory blocks (two for the membership functions and one for the numbers) are required if only two neighbored membership functions overlap. A higher overlap degree requires one more memory block.

The FPGA implementation allows the definition of 8 different membership functions for each input/output variable with a resolution of 8 bit, which requires a storage capacity of $SR = 2 * 256 * 8 + 2 * 256 * 2 = 5120$ bit $= 640$ bytes per input/output variable. If a higher resolution or more flexibility is demanded, e.g. for fuzzy processors, then a membership function generator [8] can be used.

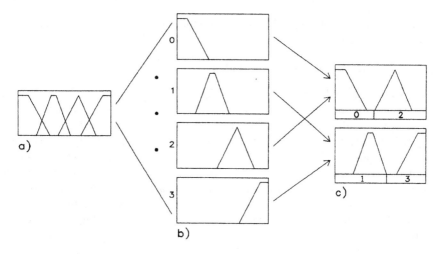

Fig. 3. Compressed storage of neighored membership functions

Figure 4 describes the calculation of the truth value with an optimized memory organization. The input value (183) is the address for the two memory blocks. There, the membership functions (MFs) 5 and 6 are activated with $\alpha_5 = 204$ and $\alpha_6 = 128$. MF 5 is selected in the subpremise of the rule base so that the multiplexer switches α_5 to 204. For each other MF a zero is processed by the multiplexer. The memory organisation with odd and even MF numbers saves an extra bit while the LSB equals one for odd binary numbers, so that instead of 3 bit only 2 bits are required.

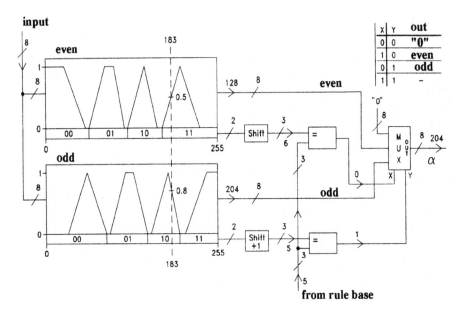

Fig. 4. Calculation of the truth value of the subpremise

3.3 Calculation of activated rules

As mentioned in chapter 2 the fuzzy logic conjunction "and" is interpreted as the minimum operation of the subpremises truth values α^i. If the truth value ω of a fuzzy rule is greater than zero then the rule is activated and delivers a contribution to the output value. The truth value ω_{act} for a fuzzy rule is stored in the register block with the MF number (Fig. 5) if ω_{act} is greater than the previous truth value ω_{pre} for the MF (maximum operation). Depending on the MF number, the multiplexer connects the truth value ω_{pre} back to the maximum circuit. After all rules are evaluated, the values of the first register block Reg0 ... Reg7 are transferred in the second register block Reg0' ... Reg7' of the second pipeline stage.

3.4 Composition rule of inference

The MFs for the output variables are organized similar to the MF for the input variables, so that the same memory organization is used. The inference algorithm limits the output MF with the truth value ω ($I_i' = \{min(\omega_i, x) \mid x \in I_i\}$). Therefore, all membership values have to be computed to calculate the resulting MF (Fig. 6). An 8 bit counter generates the addresses for the odd and even memory blocks. Together with the membership value the binary number of the membership function is selected to address the register block with the truth value ω. The minimum circuit limits the output membership function and the maximum circuit computes the compositional rule of inference. Different fuzzy

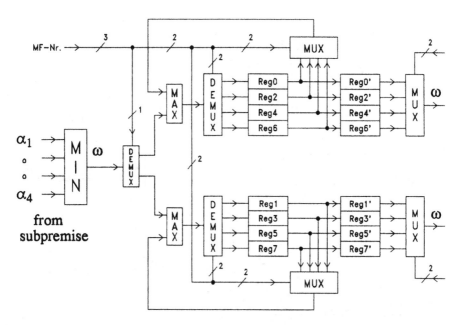

Fig. 5. Fuzzy "and" and computation of activated rules

controllers can be connected over the CAS_{in} and CAS_{out} lines to build complex systems.

3.5 Defuzzifier

To evaluate a "crisp" value (unit pulse) from the output membership function the COG has to be calculated (chap. 2). Therefore summation, multiplication and division operations have to be carried out. The concept of the repeated adder of Watanabe et al. [3] economises the multiplication operation (Fig. 7). The denominator is simply the summation of the data stream from the inference stage. Since the numerator can be computed by repeatedly adding the denominator, all summation operations can be done in the second pipeline stage. The division operation occurs in the third pipeline stage after the calculation of the numerator and denominator.

3.6 FPGA implementation

A prototype of the architecture with 4 inputs, 1 ouput and 256 rules is implemented on 2 FPGA (XC3090-100 PG84C, 320 CLBs [12]), both using about 80% of the CLBs. The rule base and membership functions are stored in external RAMs (8 bit, 20ns). The fuzzy conjunction and the computation of activated rules (Fig. 5), together with minimum and maximum circuits (Fig. 6), is placed on the first FPGA. The repeated adder is placed on the second FPGA together

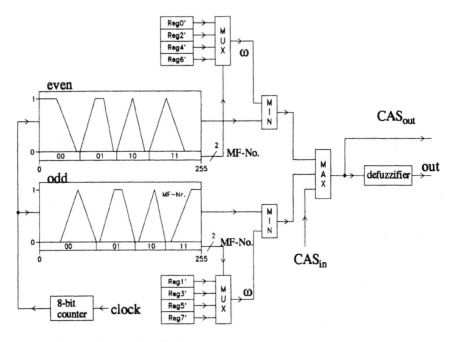

Fig. 6. Compositional rule of inference

with the divider for the defuzzification. The place and route is done half automatic which means that we placed the critical blocks and wired the critical paths by hand.

The third generation of LCA devices like the XILINX XC 4000 family can be considerably increase the performance of the described FC. By using the new devices the number of chips will also be reduced, so that the controller can easily be placed on a single FPGA.

4 Conclusion

We have introduced an optimized fuzzy controller (FC) architecture and its realization with field programmable gate arrays (FPGAs). In consideration of data dependencies and minor user restrictions by the definition of fuzzy rules (FR), it was possible to develop a high speed FPGA architecture. A prototype of the FC operates at 5MHz and needs $50\mu s$ (8 bit resolution) operation time independent of the number of inputs/outputs with 256 fuzzy rules. Future research will focus on microelectronic VLSI realization of adaptive fuzzy controllers [13] as well as different applications.

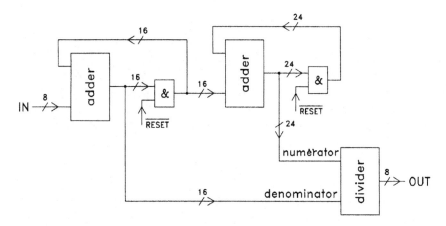

Fig. 7. Repeated adder [3]

References

1. Mamdani, E.H. : Application of fuzzy algorithms for the control of a dynamic plant. Proc. IEEE 121, No.12, (1974) 1585-1588
2. Surmann, H., Heesche, K., Hoh, H., Goser, K., Rudolf, R.: Entwicklungsumgebung für Fuzzy-Controller mit neuronaler Komponente. VDE-Fachtagung Technische Anwendungen von Fuzzy-Systemen, Dortmund, (12/13.Nov. 1992) 288-297
3. Watanabe, H., Dettloff W.D., Yount, K. : A VLSI fuzzy logic controller with re-configurable, cascadable architecture. IEEE journal of solid state circuits, Vol.25, No.2, (1990) 376-382
4. Ungering, A., Qubbaj B., Goser, K.: Geschwindigkeits- und speicheroptimierte VLSI-Architektur für Fuzzy-Controller. VDE-Fachtagung Technische Anwendungen von Fuzzy-Systemen, Dortmund, (12/13. Nov. 1992) 317-325
5. Eichfeld, H., Löhner, M., Müller, M.: Architecture of a Fuzzy Logic Controller with optimized memory organisation and operator design. Int. Conf. on Fuzzy Systems, FUZZ-IEEE '92, San Diego, (March 8-12, 1992)
6. Surmann, H., Tauber, T., Ungering, A., Goser, K.: Architecture of a Fuzzy Con-troller based on Field Programmable Gate Arrays. 2nd International Workshop on Field-Progammable Logic and Applications, Wien, 31. (Aug. - 2. Sept. 1992)
7. Surmann, H., Ungering, A., Goser, K.: Fuzzy Controller mit VLSI-Pipeline-Architektur für hohe Datenraten. ITG-Fachbericht 119, Mikroelektronik für die Informationstechnik, Stuttgart, (4-6 März 1992) 195-200
8. Ungering, A., Roer, P., Surmann, H., Daszkiewicz, D., Goser, K.: Architek-turkonzept eines Fuzzy-RISC-Prozessors mit optimierten Speicherbedarf. ITG/ GME/GI-Fachtagung, Rechnergestützter Entwurf und Architektur mikroelektro-nischer Systeme, Darmstadt, (23/24. Nov. 1992) 229-238
9. Zadeh, L.A.: Outline of a New Approach to the Analysis of Complex Systems and Decision Processes. IEEE Transactions on systems, man and cybernetics, Vol.SMC-3, No.1, (Jan. 1973) 28-32
10. Surmann, H., Flinspach, G.: Fuzzy-Controller gesteuertes Schnell-Ladeverfahren für NiCd-Akkumulatoren. VDE-Fachtagung Technische Anwendungen von Fuzzy-Systemen, Dortmund (12/13. Nov. 1992) 159-168

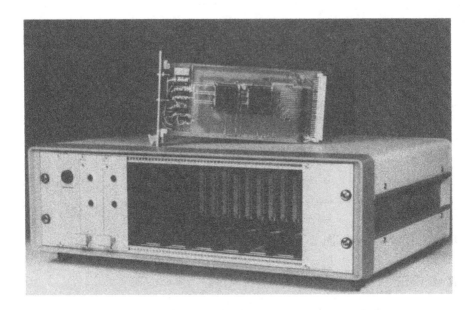

Fig. 8. Development board

11. Kosko, B.: Neural networks and fuzzy systems - A dynamical systems approach to machine intelligence. Prentice-Hall (1992)
12. XILINX Inc.: The Programmable Gate Array Data Book. Users Guide and Tutorial Book. San Jose / California, (1991)
13. Surmann, H., Möller, B., Goser, K.: A distributed self-organizing fuzzy rule-based system. Proceedings Neuro-Nimes 92 (1992) 187-194

A Real-Time Kernel - Rapid Prototyping with VHDL and FPGAs

Lennart Lindh, University of Eskilstuna/Västerås [1]

Klaus Müller-Glaser, University of Erlangen/Nürnberg [2]

Hans Rauch, University of Erlangen/Nürnberg [2]

Frank Stanischewski, University of Erlangen/Nürnberg [3]

ABSTRACT In order to remain competitive a company needs to decrease the development time for their products. Consequently the time for prototyping has to be cut down as well especially to see whether the product will fulfil requirements. One way to reduce the development time is to describe the circuit with a hardware description language, to synthesize the design description automatically and to implement in FPGAs (Field Programmable Gate Arrays).

This article describes how to implement a small real-time kernel and a very simple CPU in hardware. During the development process the hardware description language VHDL (Very high speed integrated circuit Hardware Description Language) was used for the behavioural and data flow (RTL) description, the simulation of the hardware and for synthesis to gate level. Also it describes the fitting of the design into FPGAs for rapid prototyping and experiences about different phases of the development.

KEYWORDS

Real-Time Kernel, Rapid Prototyping, VHDL, Synthesis, FPGA

1 Introduction

To decrease the time spent in the design phase is one of the most important goals for designers and project managers. One way to achieve this goal is to use the strategies of "rapid prototyping". This means for our project the use of VHDL as description language and for simulation as well as the use of FPGAs for the implementation and validation of the concept i.e. the emulation of the physical prototype and the software. The use of FPGAs instead of ASICs (Application Specific ICs) reduces implementation time from some weeks to a few days. So changes in design can be taken into account within hours or days.

The used development systems were VIEWLOGIC [6] for design entry and simulation and ACTEL [1] for routing and programming the FPGAs. The VIEWLOGIC system provides a hierachical design entry where schematics can be mixed with VHDL models. This allowed us to get a multiple view of the design.

---

[1] University of Eksilstuna/Västerås, CUS Institute, P.O.Box 11, 72103 Västerås, Sweden, FAX : +46-21-101460

[2] University of Erlangen/Nürnberg, Institute for Computer Aided Circuit Design, Cauerstr.6, 8520 Erlangen, Germany, FAX : +49-9131-858699

[3] University of Erlangen/Nürnberg, Institute of Computer Science (IMMD 3), Martensstr.3, 8520 Erlangen, Germany, FAX : +49-9131-39388

In the following chapters the way is described how we implemented our ideas of a "real" real-time processor and how we matched our specifications which originated from demands from the real-time systems area. In real-time systems there is a need in deterministic hardware and software to be able to use the systems in hard real-time applications (see chapter 2). We also tried to speed up our system by migrating software functions into hardware. This new hard- and software concept called FASTCHART (a FASt and Time deterministic Cpu and HArdware based Real-Time kernel) should make it easier to implement real-time systems.

2 A Real-Time System - A Short Description

Today it is usual to characterize a computer system with its average through-put rate. For real-time systems the average throughput rate is not the most important characteristic. Here a quotation of John Stankovic [5] is given : "A Real-Time systems is a system where the correctness of the system depends not only on the results of computations but also on the time at which the result is produced". This means that in real-time systems the most important value is time.

To determine whether a correct result is produced at the correct time the worst case execution times of all code parts in the system are needed. The worst case execution time means that time which will never be exceeded when executing the code. The code parts in a real-time system are the programs of the corresponding tasks of a multitasking system.

To ensure whether the real-time system will produce correct results at any time – in technical terms : the system is feasible – one has to correlate the worst case execution times of the task program codes with the period times of the corresponding tasks([3]). By this one gets a "virtual" utilisation of the system for every task. If the sum of all utilisations is less or equal to 100 percent the system is feasible. The deadline is that point of time at which the result of a task has to be avialable at the latest. If a deadline of one task is exceeded once the system starts to work incorrectly. Real-time systems with such hard time requirements are called "hard real-time systems". These systems are necessary in process control systems, for example in automotive or power plant applications. In literature one can also find the expression "soft real-time system". In these systems some exceedings of deadlines are tolerated, the time requirements are softer.

2.1 Overview of FASTCHART

We talked about worst case execution time of the task program code. To get the worst case time of one task one has to sum the execution times of all processor instructions in the longest (critical) program path. The problem when summa-rizing the time values is that these values can vary depending on the actions and results in the past. So one cannot say that the execution time of one processor instruction has a fixed value. One can only define the worst case time of that instruction. The differences in instruction execution times occur through the use of modern processor architecture which tries to increase the average throughput rate. Todays well known processor accelerators are pipelines or caches. These accelerators cannot generally be used in hard real-time systems.

If one takes only the worst case execution times of the processor instructions one has a great loss of performance because the real-time system is calculated on a base of a slower processor. To get deterministic execution times one has to omit pipelines and caches. Hardware interrupt and DMA (Direct Memory Acces) has to be omitted too because the processor can block when executing task code.

The Central Processing Unit CPU in the FASTCHART concept is designed to get deterministic execution times for every instruction([2]). Because of the lack of the accelerator mechanism we choose a Load-Store architecture with a fixed instruction format.

Additionally we increase the performance by implementing a task switch mechanism. This has two benefits. First we can manage a task switch (exchange the register contents) in zero time related to the CPU execution. This will give us an up to 10 percent performance increase. Second the task switch becomes deterministic because it costs no time.

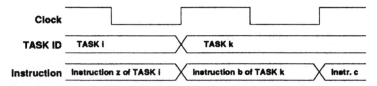

Figure 1: Task switch timing

To get a fully deterministic real-time system the operating system has to be made deterministic too. Some functions of the operating system work with wait queues where the number of items in a queue varies during run time. So the execution times of these functions will vary because they depend on the number of items in the queues, too.

Because it is impossible to build up an operating system without any wait queues we migrated the operating system kernel into hardware in a unit running concurrently to the CPU (see figure 2). This increases the overall performance of the processor and the task switch mechanism can be controlled very easily. The unit which executes the "hardware" operating system kernel is called Real Time Unit RTU.

The RTU contains a scheduler with a rate-monotonic scheduling algorithm, a dispather which controls the task switch mechanism, two wait queues one for inactive tasks and one for tasks waiting for a time event and a ready queue for tasks waiting to be executed by the CPU. Because of the partitioning of the kernel into separate parts we get a modular structure of the RTU. So we can add very easily extra units to extent the functionality of the RTU.

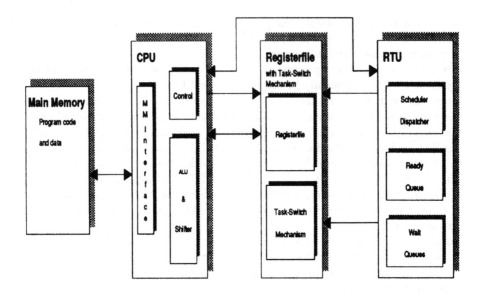

Figure 2: The FASTCHART concept overview

3 Design Overview

As mentioned above the FASTCHART concept is based on a modular structure
so the two main parts of the processor can be implemented separately. The small
interface necessary consists of one port of the register file which will be accessed
by the CPU and some control lines.

The CPU consists of an ALU, an instruction decoder and the main memory
interface. The implementation of the CPU is "RISC-like" which means that all
instructions can be executed in one CPU cycle. The instructions use a fixed
format to have an easier and faster decode. To execute one instruction the
CPU firstly accesses the register file to get the instruction and the new program
counter. The instruction is decoded and executed by using the ALU or the main
memory interface. Afterwards the new instruction is fetched and stored in the
register file. Also the program counter is calculated newly and stored. If the
instruction decoder detects an instruction which controls the RTU the CPU sets
the corresponding control lines and executes a NO-operation by itself.

The RTU is implemented by using the "black box" concept where the functions
of the RTU are divided into three black boxes. The most important black box
is the Scheduler/Ready Queue Box. It has an interface to the Register File, the
Wait Box and the Terminate Box. The Scheduler/Ready Queue Box outputs
the ID of the running task which is taken to select the correct register contents
in the Register File. The task IDs from the other two boxes are the input. This
input of task IDs represents the changing of the status of the tasks. The status

of a task can change from inactive or waiting to the ready status by detecting a time event or by activation by another task.

The Scheduler/Ready Queue Box consists of a small control unit, a priority decoder and some FIFOs. The Ready Queue is represented by the FIFOs which store the ready task IDs. For every priority in the system there exists one FIFO. Because of the limitation of the hardware the maximum entry count of the FIFOs is fixed that means that there can exist as many tasks of the same priority as the maximum entry count of the corresponding FIFO defines. To detect the ready task with the highest priority the Priority Decoder looks for that FIFO which is not empty and which has the highest priority of all not empty FIFOs. The priority of this FIFO is then the priority of the highest ready task. If this priority is higher than the priority of the running task a task switch is initiated.

The black box Wait Queue is used to take up tasks which delay for a specified time. The Wait Queue box outputs task IDs to the Ready Queue if the tasks have reached their delay time. The Wait Queue Box consists internally of a counter for every task which is decremented by every system time tick. If a counter has reached zero the Wait Queue controlling unit initiates a transfer of the corresponding task ID. The input of the Wait Queue Box gets a task ID and a delay time whereby the counter of that task is set with the delay time.

Because it is not possible to create and destroy tasks dynamically in the hardware implementation the maximum number of tasks administered by the RTU is fixed. Also a task ID in RTU cannot vanish in the system. So the separate unit is needed to hold all IDs of inactive tasks. This job is done by the Terminate Box. It outputs a task ID if this task is activated by another task and transfers it to the Ready Queue. The Terminate Box is composed of a small RAM where the status of every task if stored. If a task is activated the CPU sets the Activate Control Line and sends the task number to the Terminate Box. The Terminate Box toggles the status of this task and transfers the ID. If a task terminates the Terminate Control Line is set and it causes a toggle of the status only.

If a task executes an RTU command which causes a task switch its task ID is taken and send to the corresponding black box to hold it and perform the corresponding operation. Only if a task is preempted that means there is a ready task with higher priority which takes the CPU now the task ID of the running task is not sent to the other black boxes but transfered directly to the Ready Queue.

4 Design History

All items discussed in the chapter "Design Overview" were first of all only un-proved ideas. So we decided to implement a small prototype to prove if our ideas work in reality. For the implementation we chose the way of rapid prototyping to quickly get an answer. When using rapid prototyping it is useful to structure the way for the implementation. We chose the "Top Down" method. For our example we got six design steps:

- Step 1: Specification and system design
- Step 2: Functional description of the subsystems
- Step 3: Integration and simulation of all modules
- Step 4: Technology mapping and first timing analysis
- Step 5: Routing, optimization and accurate timing analysis
- Step 6: Construction and test of the prototype

In step 1 we specified the functionality of the real-time processor prototype. This means firstly we had to specify the instruction set and following from that the functions of the CPU and RTU. After the specification we decomposed the functions into subsystems to get a better structured design. During the decomposition we also defined the interfaces between the subsystems. All this work was done as the only step in the implementation without computer help. But what we needed was a detailed knowledge about the potentialities of the hardware components which we took for the implementation to know at which granularity we had to divide the processor functions into the subsystems. After the specification and the decomposition we wrote the first test programs before we thought about the implementation to get no interference with the things we wanted to test and the things we could test.

In the description step (Step 2) we described the behaviour of the subsystems. Therefore we took the hardware description language VHDL to have the possibility of describing the behaviour without thinking directly in hardware. For example we show here the VHDL code of a FIFO which is used in the ready queue in the RTU. As one can see we only then need knowledge about the hardware when we write program parts which access the interface of this subsystem or which interact the magnitude of bits we want to store in the FIFO.

```
-- Description of the interface of the FIFO subsystem
entity FIFO is
 port(
      signal clk, reset :  in bit;
      signal push_task_id, pop_task_id :  in bit;
      signal empty :  out bit;
      signal data_task_id_in :   in vlbit_1d(2 down to 0);
      signal data_task_id_out :   out vlbit_1d(2 downto 0)
     );
end FIFO;

-- Description of the behaviour of the FIFO subsystem
architecture FIFO_BEH of FIFO is

:process
begin
 wait until clk = '1';

 if push_task_id = '1' then
```

```
  temp := addum(last_p,one);
  last_p(2 downto 0) <= temp(2 downto 0);
 elsif pop_task_id = '1' then
  temp := addum(next_p,one);
 endif;
 next_p(2 downto 0) <= temp(2 downto 0);
end process;

:process(last_pointer, next_pointer)
begin
 if last_pointer = next_pointer then
  empty <= '1';
 else
  empty <= '0';
 end if;
end process;

end FIFO_BEHV;
```

In step 3 we began with the separate simulation of the subsystems to see if their behaviour fitted into our specifications. Then we connected all subsystems to get the complete processor. Therefore we used the possibilites of the VIEWlogic design system where we can mix schematics and VHDL modells in the same design. We made the system simulation where we looked for the functionality of the prototype and not for the performance. The simulation models were the same as we used for the description of the behaviour so we did not need to write new code. This has the advantage that we do not get new sources of errors in our implementation.

```
begin INIT_TASK               begin TASK_1
   ACTIVATE(TASK_1,PR_0);      start:  DELAY(16 cpu_clocks);
   STARTADDRESS(20H);               JMP start;
   ACTIVATE(TASK_2,PR_1);      end TASK_1;
   STARTADDRESS(40H);
   ACTIVATE(TASK_3,PR_0);      begin TASK_2
   STARTADDRESS(60H);          start:  DELAY(16 cpu_clocks);
   TERMINATE;                       JMP start;
end INIT_TASK;                 end TASK_2

                               begin TASK_3
                               start:  DELAY(16 cpu_clocks);
                                    JMP start;
                               end TASK_3;
```

To verify our desing we executed our test programs by the simulated prototype. In the following we list one of the test programs as an example to show how we tested the subsystems of the prototype. The example program tests all parts of the RTU : the Terminate Box, Wait Queue Box and the Scheduler/Ready Queue Box. Therefore we have four tasks which are separate and sequential programs.

The first task (INIT_TASK) initializes and starts the other tasks with different priorities like in a startup phase. After the startup phase this task terminates. The other tasks (TASK_1, TASK_2, TASK_3) are all periodic that means they run in a loop in which a delay instruction is located. When executing the delay instruction the task will be stopped, taken from the CPU and transfered to the Wait Queue. After delaying for the specified time the task is transfered to the Ready Queue and will compete for the CPU. The Scheduler will detect the task with the highest priority in the Ready Queue and will start the execution of this task by the CPU if the priority is higher than the priority of the running one.

In the first three steps we had no contact with the "real" hardware. We simulated the behaviour of the processor prototype on an abstract level with no knowledge of the underlying hardware. Then in step 4 we made the first technology mapping by synthesis of the VHDL models and got the logic structure at gate level. Because we wanted to use FPGAs from the ACTEL company we used the A1000 library for the synthesis. The synthesis tool was part of the development tool of VIEWlogic so we could also directly use the gate level models in our mixed level simulation.

During the synthesis we decided how many FPGAs it would take. It was like an iterative process because first one had to synthesize the VHDL models to get a feeling about the number of gates needed. We got to know that it was not possible to fit the whole specified functionality of our system into one single ACTEL FPGA. After the device fitting, an interactive process of decomposition and composition, we saw that we needed 6 FPGAs. After decomposition we tried to fit the resulting subsystems into one FPGA considering not only the number of gates available and needed but also the number of input/output lines of the devices. Only if a subsystem could not be fit to one FPGA device we had to decompose once more. The composition of modules of the system was done with respect to get a stucture with short ways between corresponding modules. Another point was to minimize the number of FPGA devices by reaching a high utilisation rate of every device.

With the VIEWlogic development tool we could immediatly simulate the new composition of the system. The delay time for the building blocks in the synthesized models was first set to 10 nanoseconds. It was impossible to consider the real delay times influenced by the capacitive load (number of fanins and wire length) in this phase. This simulation gives only an impression of the performance of the designed system.

In step 5 we made a full technology mapping by routing all FPGAs. The routing is needed first to get the real delay informations and second to get the connection list for the internal wiring needed to program the FPGAs. The routing program from ACTEL checks the netlists for errors like asynchronous feedbacks. Then it optimizes the design in terms of the minimum number of building blocks and to reduce the length of the internal connection wires. After optimization the connection list for the programming purpose and informations of the real delay are avialable.

After successful routing a real delay time simulation is possible. If this simulation carries out new time errors the design process has to be continued by starting step

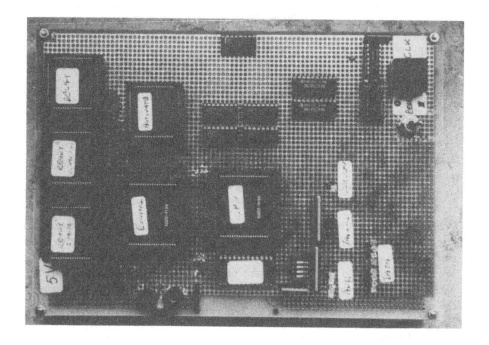

Figure 3: Task Photograph of the FASTCHART prototype

4, decomposition and composition, again. To minimize the number of these very time consuming simulation runs needed for timing verification it is important to find an optimum set of stimuli.

Step 6 in our design flow was the fastest one. Here we constructed the prototype and made an update of all development documents. The program of the FPGA devices was started after successful system simulation. We developed a prototype board (size 6U) in wire-wrap technology and sockets. After the test of the wiring of the board we mounted all FPGAs and other devices (EPROM, RAM, Drivers, LEDs) necessary to get the whole functionality of the system. We loaded our test programs into the EPROM device for the test of the developed prototype.

5 Results

Figure 3 shows a photograph of the developed FASTCHART prototype containing the EPROM devices, four RAM devices, two driver devices, a clock generator, resistor networks and LEDs and two ACTEL PGAs type A1020 and four ACTEL FPGAs type A1010.

Table 1 shows the time durations of the different design phases. After a concept phase of 6 weeks the implementation took another 8 weeks. It should be pointed out that the phase of the "real" implementation, step 6, with the programming of the FPGAs, the wrapping of the wires and the system test was carried out within one single week.

Design Phase	Duration
Concept phase	6 weeks
Implementation :	
- step 1 to 3	5 weeks
- step 4 to 5	2 weeks
- step 6	1 week
Sum	14 weeks

Table 1: Duration of the different design phases

When using rapid prototyping one unfortunately has to make concessions. In our prototype we reduced the functionality of the CPU and RTU only to show the principle of the real-time processor. Another point is that the speed of our prototype is not too high, the system clock rate is about 500 kHz. But the performance of our prototype is good enough for emulation of the system with the advantage of a big reduction in time spend for software/hardware codesign. Because the time for emulation is magnitudes shorter than the time necessary for simulation.

We did not spend too much time to get a utilisation rate higher than 70 to 80 percent. During the development process we got to know that the hardware requirements have to be considered earlier than in step 4. Additional time to get a higher utilisation rate would give the result of a successful reduction of the number of FPGAs by one.

By using the technology of these FPGAs we had to consider that there is only one clock signal and there are no internal tri-state-buffers avialable. This means that some lower functionality has to be carried out by more complex logic structures that means with more building blocks.

We made some helpful experiences when using VHDL synthesis in this development and some other projects [4]: As systems can be modelled on different levels of description VHDL can be used for a large range of steps within the ASIC development process. VHDL models written for the purpose of simulation or documentation can be exchanged between different systems according to the full VHDL-1076 standard. For VHDL-synthesis it is important that only a subset of VHDL can be used for writing models. To get synthesizable VHDL code every synthesis tool requires an own special style of modelling.

The performance of the netlist generated by synthesis can be controlled by means of modifying the VHDL model or by application of the systhesis constraints provided by the synthesis tool. Usually synthesis contraints are well documented. But it is not easy to see which modelling style leads to the best synthesis result.

Synthesis tools of today convert a real behavioural VHDL description to a netlist. Synthesis needs some information about the association of operation to certain time cycles and about the block structure. Based on a tool specific synthesizable RTL description with VHDL synthesis will provide a technology dependent netlist on gate level according to the given synthesis contraints. Special conditions such as design for testability have to be observed during synthesis. Test pattern generation has to be done in parallel to the main design flow.

There are several possible ways to describe a given functionality. One can select for instance between a more abstract and a more structural view, between concurrent amd sequential processes and "if-then-else"- and "case-when-end"- statements. Busses can be modelled as vectors or single signals used in loops.

Further experience shows :

- The number of operations occuring in a VHDL model has a great influence on the synthesis result, because redundant operators cannot be eliminated completely in the optimization phase.

- VHDL descriptions written in a more structural style show better synthesis results. But there are some limitations for the optimization of speed and area. Abstract VHDL descriptions usually save design time without decreasing the performance too much.

- Different VHDL statements with identical functionality do not influence the synthesis results.

- To model busses one should use vectors.

- A strictly delay constraint synthesis takes more computing time than an area limited synthesis run.

6   Conclusions

The use of VHDL with its possbility to abstract during the behaviour description to a higher level and the capability to synthesize the abstract described functionality to hardware allows the designer to be more productive. So it is possible to handle higher design complexities.

The use of FPGAs shortens the time after the system verification phase. The time to get a prototype or to make a redesign is much shorter here compared with the time necessary to get the corresponding ASIC. The possibility to emulate a system with an FPGA prototype showed earlier errors in design and specification not recognised till now. It allows an early test of cooperation of the software and the hardware (soft- and hardware codesign).

We also see some additional factors which will decrease the development time. One important factor is one easy to use development tool which integrates all design tools and has good interfaces to the other needed products. Like in the VIEWlogic product one can in one describe and simulate all parts of the design even when different design levels like VHDL or gate level model are used. Also it is important that one can simulate the whole design without implementing any hardware. So one can find the errors in the design and can change it immediatly. The results one gets with rapid prototyping are not the optimum solution but they are good enough to show that the principle will work. As in our case we can show that the migration of software into hardware will work and the demands on the hardware are not extraordinary.

The work was carried out at the Institute for Computer Aided Circuit Design and the Institute of Computer Science, Department of Computer Structures, University of Erlangen-Nurnberg, Germany. And it was supported by the Fraunhofer Institute of Integrated Circuits, Erlangen, Germany and was sponsered by the Swedish National Board for Technical Development (STU today NUTEK) and the Bavarian Government.

References

[1] Actel Corporation, 955 East Arques Avenue Sunnyvale, CA 94086, USA

[2] Lindh, Lennart; Stanischewski, Frank; "FASTCHART - A Fast Time Deterministic CPU and Hardware Based Real-Time-Kernel"; Proceedings of Euromicro Workshop on Real-Time Systems Paris 1991; IEEE; pages 36-40;

[3] Liu, C.L.; Layland, James W.; "Scheduling Algorithms for Multiprogramming in a Hard-Real-Time Environment"; Journal of ACM; 1973; Vol. 20; No. 1; pages 46-61;

[4] Selz, M., Bartels, S., Syassen, J.; "Synthesis with VHDL"; Procedings of the VHDL Forum 1993; Maerz 1993; pages 31-40;

[5] Stancovic, J.A.; "Misconceptions about Real-Time Computing"; IEEE Comp; No. 21; 1988;

[6] Viewlogic Systems, Inc. 293 Boston Post Road West Marlboro, Ma 01752-4615, USA

JAPROC - A 8 bit Micro Controller Design and its Test Environment

Herbert Grünbacher
Alexander Jaud

Institut für Technische Informatik
Vienna University of Technology

Treitlstr 3/182.2
Vienna/Austria

Tel (+43.1) 588 01-8150
Fax (+43.1) 56 96 97
e-mail herbert@vlsivie.tuwien.ac.at

Abstract. This paper describes the design of JAPROC, an 8-bit micro controller. JAPROC is a processor-core which is being developed within the EUREKA project JAMIE. The design consists of approximately 5000 gates and has been implemented in a FPGA Xilinx X4005.

The design serves as a prototype for a full custom processor-core for smart card applications.

For testing purposes a PC board has been developed which allows to configure the FPGA, download and execute micro controller code and compare the results to an emulator.

1 INTRODUCTION

JAPROC is an 8-bit micro controller prototype for full custom implementation later on. The full custom design will eventually serve as a processor-core in smart card application. JAPROC is being developed within the EUREKA project JAMIE [1].

The main purpose of this design is to verify whether the instruction set is implemented properly. For compatibility reasons the instruction set should be upwards compatible to the PIC instruction set [2]. There is no strict timing requirements, but the program behavior of the processor needs to match to a given PC based reference emulator.

It is expected that the micro controller needs to be modified for particular smart card applications. The FPGA design should then serve as a vehicle to verify the design changes.

A PC board has been developed which contains all AT bus logic as well as the FPGA as such. The FPGA can be programmed by downloading the personalization pattern from the PC to the FPGA board. A binary representation of the emulator input can be downloaded and executed in the micro controller.

By comparing the results of the FPGA micro controller to the corresponding results of the emulator the micro controller design can be verified.

2 ARCHITECTURE

The design is based on a load/store architecture, busses and memories for data and instructions are separated (Harvard architecture). This architecture guarantees that a faulty program can't overwrite the content of the instruction memory.

Data bus and memory are 8 bit wide, while program bus and memory have a width of 12 bits.

The register file is composed of 32 addressable 8 bit registers including the I/O ports, status register, program counter (up to 12 bits) and real time clock/counter register.

FILE ADRESS	REGISTER	COMMENT
00	INDIRECT ADDR.	
01	RTCC	Real time clock/counter reg.
02	PC	12 bits wide program counter
03	STATUS	
04	FSR	Bits 0-4 select one of the 32 available file registers in the indirect addressing mode (that is, calling for file f0 in any instruction)
05	PORT A	4 Bit I/O register
06	PORT B	8 Bit I/O register
07	PORT C	8 Bit I/O register
08..1F	GENERAL PURPOSE	

Fig 1. Register File Arrangement

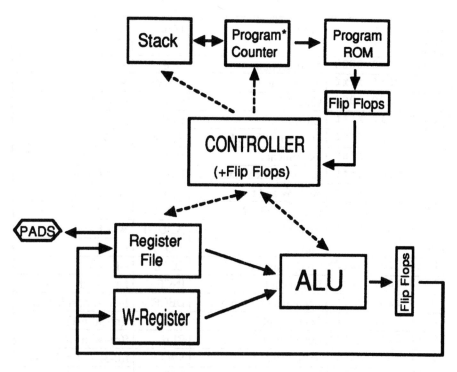

Fig. 2.   Block Diagram of the Micro Controller

The 8-bit ALU contains one additional temporary register (W register) and gating to perform Boolean functions, bit manipulation and shift operations between data held in the W register and any file register. The arithmetic status of the ALU, carry bit and zero bit, is held in the status register.
An on-chip two-level stack is employed to provide easy to use subroutine nesting.

3   INSTRUCTION SET

Each of the 30 instructions is a 12-bit word divided into an OP code which specifies the instruction type and one or more operands which further specify the operation of the instruction. All instructions are executed within one single instruction cycle, unless a conditional test is true or the program counter is changed as a result of an instruction. In this case, the execution takes two instruction cycles.
The instruction set allows bit, byte and register operations. Data can be addressed direct, or indirect using the file select register. Immediate data addressing is supported by special "literal" instructions which load data from program memory into the W register. Program control operations, supporting direct, indirect, relative addressing modes, can be performed by Bit Test and Skip instructions, Call instructions, Jump instructions or by loading computed addresses into the PC.

Instruction Binary	Name	Mnem.	Operation
0000 0000 0000	No operation	NOP	
0000 001f ffff	Move W to f	MOVWF	W → f
0000 0100 0000	Clear W	CLRW	0 → W
0000 011f ffff	Clear f	CLRF	0 → f
0000 10df ffff	Subtract W from f	SUBWF	f - W → d
0000 11df ffff	Decrement f	DECF	f - 1 → d
0001 00df ffff	Inclusive OR W and f	IORWF	W ∨ f → d
0001 01df ffff	AND W and f	ANDWF	W ∧ f → d
0001 10df ffff	Exclusive OR W and f	XORWF	W ⊕ f → d
0001 11df ffff	Add W and f	ADDWF	W + f → d
0010 00df ffff	Move f	MOVF	f → d
0010 01df ffff	Complement f	COMF	¬f → d
0010 10df ffff	Increment f	INCF	f + 1 → d
0010 11df ffff	Decrement f, Skip if 0	DECFSZ	f - 1 → d, skip if zero
0011 00df ffff	Rotate right f	RRF	f(n)→d(n-1)
0011 01df ffff	Rotate left f	RLF	f(n)→d(n+1)
0011 10df ffff	Swap halves f	SWAPF	f(0-3) ↔ f(4-7) → d
0011 11df ffff	Increment f, Skip if 0	INCFSZ	f + 1 → d, skip if 0
0100 bbbf ffff	Bit Clear f	BCF	0 → f(b)
0101 bbbf ffff	Bit Set f	BSF	1 → f(b)
0110 bbbf ffff	Bit Test f, Skip if Clear	BTFSC	Test bit(b): Skip if 0
0111 bbbf ffff	Bit Test f, Skip if Set	BTFSS	Test bit(b): Skip if 1
1000 kkkk kkkk	Return, Literal in W	RETLW	k → W, Stack → PC
1001 kkkk kkkk	Call subroutine	CALL	PC+1→Stack, k→PC
101k kkkk kkkk	Go to address	GOTO	k → PC
1100 kkkk kkkk	Move Literal to W	MOVLW	k → W
1101 kkkk kkkk	Incl. OR Literal and W	IORLW	k ∨ W → W
1110 kkkk kkkk	AND Literal and W	ANDLW	k ∧ W → W
1111 kkkk kkkk	Excl. OR Literal and W	XORLW	k ⊕ W → W

Fig. 4. Implemented Instruction Set

4 IMPLEMENTATION

The design of the micro controller has been carried out on Sun-Sparc compatible workstations using Viewlogic and Xilinx software [4].

The target FPGA for the micro controller including instruction ROM and gating for test purposes is a Xilinx X4005-6.

We implemented a 32 word only ROM as instruction-memory on chip for testing purposes. The size was limited by the number of unused CLBs after implementing all the other parts. This ROM can easily be extended by adding external hardware to the FPGA, as there are 30 I/O pins unused. The ROM has been implemented by using Xilinx Configurable Logic Blocks (CLBs) as ROM.

The register file has been implemented by using a 32x8 register (RAM) macro for area efficiency. The first 8 registers of this macro needed to be duplicated due to particular read/write requirements of the register operations. These eight registers and the W register and the stack is build up by flip-flops.

The controller logic has been optimized, technology mapping has been done manually to better utilize the CLB structure.

The ALU has more functions (a total of 32) then presently required (17) to allow easy modification of the instruction set.

The utilization of the X4005-6 is as follows (extracted from the report file of the PPR):

- Equivalent "Gate Array" Gates: 4300
- Used packed CLBs: 172 (out of 196)
- Flip Flops: 161 (out of 504)
- 3-State Buffers: 53   (out of 504)

30% of the used CLBs are occupied by the ALU gating, another 30% are needed by the controller, 10% by the program counter, 13% by the register file.

Normally transparent latches are used to implement a pipelined design. But this would use up a lot of the configurable logic of a Xilinx FPGA since latches have to build up by using this logic blocks. So we separated the pipeline-stages by flip-flops. Another implementation driven constraint refers to the special purpose registers. These are also made of flip-flops because it's not possible to configure RAM blocks smaller than 16 bits. But there are only eight special purpose registers.

5 TEST ENVIRONMENT

For cost reasons all testing is done in a PC environment [3].

A test board contains sockets for two FPGAs and the logic to interface the board to the AT bus.

The FPGA personalization bit stream can be downloaded from PC under program control.

The test patterns are generated by writing assembler programs and running them on the PC-emulator of the micro controller. The emulator stimuli are converted to machine code and can be downloaded to the FPGA board and run as program for the micro controller. The micro controller responses can be transmitted back via the AT bus and can than be compared with the emulator responses.

This allows a rather comfortable debugging environment.

By running emulator and real hardware in single step mode and displaying the results in two PC-windows one can easily compare results and easily detect design errors. This test environment is very flexible and test pattern can be generated at a relatively high level.

6   CONCLUSIONS

By using FPGA technology and a flexible test environment a prototype implementation of a 8-bit micro controller can be designed and debugged rather quickly. The internal logic of such FPGAs serves well for pipelined processer-core implementations.

Modifications to the instruction set or extensions to the architecture to support particular applications can be implemented and debugged easily.

7   REFERENCES

Joint Analog Microsystems Initiative of Europe (JAMIE), Technology Independent Semi-Custom/Custom ASIC Concept, mikron, 1991.

PIC 16C5x Series, Microchip Technology Inc.

Huber E., PC-Einsteckkarte und Entwurfsumgebung für Xilinx FPGAs, Internal Report 92-08.

The Programmable Gate Array Data Book, Xilinx Inc., San Jose, 1991.

Chameleon: A Workstation of a Different Colour

Beat Heeb, Cuno Pfister

Institut für Computersysteme, ETH Zürich
CH-8092 Zürich, Switzerland

Oberon microsystems Inc.
CH-4000 Basel, Switzerland

heeb@inf.ethz.ch
pfister@inf.ethz.ch

Abstract. Chameleon is an experimental workstation based on a RISC processor. It provides unprecedented flexibility and speed for certain applications due to the use of RAM-configurable Field Programmable Gate Arrays (FPGAs). FPGAs are used to replace glue logic as well as to provide a non-dedicated computation resource. This resource can be regarded as a *general purpose coprocessor* which can be reconfigured and thus transformed into a special purpose coprocessor in milliseconds at run-time. The coprocessor can be used both for handling complex input/output functions as well as to replace time-critical inner loops of user programs running on the central processing unit. Chameleon radically relies on FPGAs for all input/output functions. It serves as a means to probe the limits of FPGA usage while at the same time being the development system for its own FPGA circuits.

1 Introduction

Compared to software development, hardware development is a cumbersome activity. The result of a hardware project, a board or a chip, usually cannot be fabricated in-house. Instead, the manufacturing data has to be sent to an appropriate company, which results in high cost, long turn-around time and generally in much inconvenience. Fortunately, the availability of field-programmable gate arrays is beginning to change this situation. An FPGA can be programmed in-house, possibly even in its target board.

Field-programmable gate arrays can be used in many areas where conventional gate arrays would be used otherwise. Today, this is mainly the replacement of *glue logic*; to reduce board area and power consumption. Another area where gate arrays have been used is the implementation of *general purpose processors*. Here FPGAs are not competitive, since their programmability makes them inherently inferior in speed and density, and because their production costs are too high for such high-volume circuits.

There are some computationally expensive algorithms whose critical inner loops could be sped up by at least partially implementing them in hardware, instead of interpreting them on general purpose processors. Code outside of the inner loops is less time-critical and can be implemented in any conventional instruction set. This approach is taken with the so-called signal processors, which are microcontrollers enhanced with instructions which calculate check-sums, which reverse bits in a word (for the fast fourier transform), which combine multiply with add operations, etc. But more often than not, algorithms are too specialised to make the development of special purpose processors cost-effective. Obviously, *special purpose processors* are an area where FPGAs present new opportunities.

Since typically only a few specialised instructions are necessary, it is advantageous to use a commodity microprocessor and only implement the special instructions in an FPGA, i.e. to implement a *special purpose coprocessor*.

When we restrict ourselves to the use of RAM-programmable FPGAs, where the configuration is stored in a static RAM that can be altered anytime, it even becomes possible to build a *general purpose coprocessor*, which is turned into a special purpose coprocessor by the central processing unit configuring it.

There are limits to the usefulness of such a scheme, in particular because data have to be transfered to and from the coprocessor. If the speed of this data traffic is the limiting factor, it is faster and simpler to do all computations in the central processing unit.

To gain an understanding of how useful the concept of a general purpose coprocessor is in practice, we designed a workstation around this concept. We went so far as to delegate all input/output functions to FPGAs, which made it possible to build the very simple yet flexible workstation *Chameleon* in a very short time.

Chameleon is a complete workstation with mouse, keyboard, colour video, network and disk interface. It is not only the target system for FPGA circuits but also the development system where these circuits are specified, implemented and tested.

Other work in this area is done by a group at DEC PRL [2].

2 The Hardware

Essentially, Chameleon consists of a CPU, memory and FPGAs. The number of support chips was to be minimised, which has been the main reason why we chose the LR33000 microcontroller from LSI Logic as CPU [6]. It is a fast 32-Bit RISC processor compatible with the MIPS R3000A. It contains instruction and data caches and a DRAM-controller, which makes for a low chip-count system. On the other hand it lacks the usual floating-point unit and the memory management unit. The lack of the latter has been made up for to a degree by providing a memory protection RAM, which lets the address space be protected in 32 KB pages. Outstanding floating-point performance is not necessary for our purposes, a software FPU emulator is sufficient (however, the pin-compatible LR33050 with integrated FPU has been announced meanwhile). Integer performance is comparable to commercial workstations (> 45'000 Dhrystones at 40 MHz). Four single inline memory modules (SIMMs) yield a maximum of 64 MB of DRAM, additionally there is 1 MB of Video RAM and an EPROM providing at most 1MB of non-volatile storage.

Each of the 1152 x 910 pixels is represented by one Byte of video memory. During screen refresh, each pixel value is translated by a colour palette chip to a three–Byte RGB value. The pixel data is serially read out of the eight Video RAM chips. The DRAM–controller is requested every 12 microseconds to feed the start address of the next 256 pixels to the Video RAMs. This video refresh task is performed by an Algotronix CAL1024 FPGA [1], the so–called Control CAL. The screen resolution could be changed by reprogramming the Control CAL, e.g. to drive a lower resolution projection device.

Apart from video timing, the Control CAL is used for a variety of other tasks. In the current standard configuration, three serial receivers are used for mouse, keyboard and V24 file transfer. The global clock signals for the CAL Array are generated in the Control CAL also.

The CAL Array is a 2 by 3 matrix of CAL1024 parts, yielding a 64 by 96 array of CAL (Configurable Array Logic) cells, which are arbitrary two–input one–output functions or latches. Each CAL chip can be read and written via Byte–wide random–access to its configuration RAM, which can simply be treated as conventional static RAM. Some of the peripheral cells of the array are connected to the system bus as a 16 Bit memory–mapped port. Other peripheral cells are connected to a variety of connectors, for parallel and serial communication. There is also a memory card connector. SRAM memory cards may be used as backup medium, but may also be treated as local memory for the CAL Array.

The Chameleon board is a six–layer board roughly of DIN A4 letter size. When the CAL Array is not used, it is configured such that it consumes practically no power. This is possible because the power consumption of CMOS circuits like the CAL1024 is proportional to the number of state transitions per time unit. Thus it is sufficient to stop the clocks for an idle CAL Array and to provide a configuration which contains no oscillators. In an idle state, the machine uses about 8 Watts of power. This made it possible to dispense with a fan.

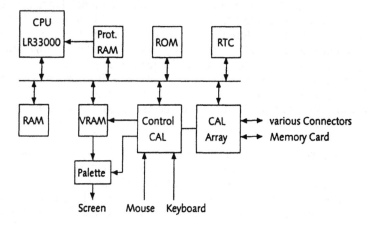

Fig. 1. Block Diagram

3 The Software

Oberon is the name of both a programming language [8] and an operating system [9]. Both have been ported to a variety of workstations, one of them based on the MIPS R3000A processor. The latter system has been adapted for Chameleon. Oberon is particularly well suited for rapidly developing interactive programs, due to its support of modular and extensible software systems.

After Oberon was up and running, the CAL development tools have been ported to Chameleon. They consist of a compiler for the hardware description language Debora [5], a layout editor with integrated autorouting [7], a CAL device driver and graphical debuggers.

Debora is a simple hardware description language which has been used to design printed circuit boards, PALs, EPLDs and FPGA configurations. A Debora text specifies the behaviour of a circuit in the form of invariants and assignments. Invariants like $Z := A + B$ or $T := U = V$ represent combinatorial logic, while assignments like $R := A$ or $S := S + 1$ represent sequential logic (Z, A, B, T, U, V, A are signals, R and S are registers). Signals can be tied together by an assortment of logical, Boolean and arithmetic operators.

Debora descriptions can be organized into structures and units (units are instances of structures), and into modules which contain structures and are compiled separately. A structure typically implements a state machine or a functional block of a larger system.

Invariants and assignments can be guarded using an IF construct. This allows to express non-determinism in the data domain, i.e. the result of an input combination which can never occur need not be specified. This is important to allow optimal implementations for different device architectures.

A register keeps its state except when a clock signal changes. The clock signal for a particular register is specified in the register's declaration.

Except for the IF statement, there are no traditional language constructs defining a particular control flow; like WHILE or FOR loops. All statements are executed in parallel, therefore the textual order of statements is not relevant. The IF construct can be used to guard assignments by a state configuration, i.e. an assignment is only executed when a Boolean condition holds. This allows to realise sequential composition of statements as well as any other control flow strategy, e.g. the transitions of a particular state machine.

Semantically, Debora could be regarded as a subset of the Unity notation [3]. However, Debora additionally allows to express asynchronous circuits.

A new logic synthesis algorithm was developed for the minimisation of CAL circuits [5]. This algorithm tries to retain as much structure of the original Debora description as possible, to ease the task of cell placement.

Cells are placed automatically using a clustering algorithm. The algorithm produces an initial layout which can be modified interactively. CALLAS [7] is a physical design framework for CAL circuits which contains a layout editor for this purpose. Special properties of this editor are the enforcement of the layout's consistency with the original Debora description ("layouts are correct by design"), and its extensibility, thus the term "framework". New layout tools can be designed, implemented and run without changing the CALLAS core system. In fact, its source code need not even be available. The easy integration of new tools allows the user to implement layout algorithms tailored to his particular needs with little effort. Having available a number of layout tools, the user can then decide on the degree of interactivity (vs. automatic placement) most appropriate for a given circuit. If the FPGA resources are scarce, the layout must be nearly optimal and thus the user will have to work mostly manually. For uncritical circuits, the layout may even be generated completely automatically.

Apart from a maze router, a clustering placement algorithm and several placement optimisation algorithms, CALLAS provides as standard facilities a graphical debugger which allows to graphically watch the state of a CAL chip's cells while they are active.

4 A Character Thinning Coprocessor

4.1 Introduction

Character thinning is a fundamental and time consuming preprocessing step in many character recognition algorithms. During thinning the strokes and curves in a rastered representation of an image are reduced to thin representations (Figure 2). The goal is to delete as many pixels as possible without changing the connectivity of the image. Many software oriented (sequential) as well as hardware oriented (parallel) algorithms were published for that problem. The algorithm used in this example is derived from [4]. To reach the desired goal, it uses several iterations, each deleting a set of pixels from the image. Similar to the Game of Life problem, the 8 immediate neighbours are used for the decision whether to delete a pixel or not. Slightly different functions are necessary for even and odd iterations to be correct in all situations. The algorithm terminates if no more pixels can be deleted during an odd and an even iteration.

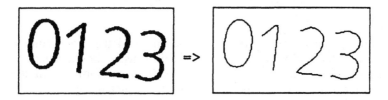

Fig. 2. Character Thinning Example

4.2 Implementation

Rastered images for thinning are stored efficiently with one bit per pixel. It is therefore desirable to process multiple pixels in parallel. In this example the new pixels are calculated row–wise one byte at a time. After initialization at the beginning of a row, three bytes (one from the actual and two from the adjacent rows) have to be written before the next byte can be read back. These four natural cycles (three writes and one read) can be used to serialize the calculation using only two function units for an eight bit result without performance loss. Arguments as well as results are therefore kept in shift registers shifting by two bits for each processor access (Figure 3).

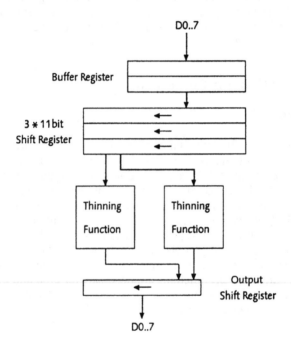

Fig. 3. Character Thinner Block Diagram

Two buffer registers are needed to temporarily hold the new argument bytes, because they have to be transferred to the shift register simultaneously. The two identical function units are directly derived from the formulas in the paper mentioned above, but could easily be adapted for a different algorithm.

An additional one bit register, which has to be set by software at the beginning of an iteration, is cleared automatically as soon as any pixel changes its state. At the end of the iteration the register can be read back to determine whether the image has reached its final state.

4.3 Speed Comparisons

Hardware solutions like the one presented, are only justified if there is a considerable performance improvement compared to a software solution. If the same algorithm, which largely depends on logical operations on individual bits, is implemented in software, it would obviously be several orders of magnitude slower. Comparisons with an algorithm optimized for a software implementation result in a factor of only 4. This discouraging result can be commented in two ways: First the hardware algorithm can be extended to 16 or 32 bit words speeding it up by an additional factor of two or four. Second, and even more important, the software part of the hardware solution can also be improved drastically. In a simple solution the whole image is fed through the thinning processor for every iteration, spending most of the time on areas not changing at all or only during the first few iterations. A more sophisticated solution with a more local control of the iteration count could therefore easily surpass a software solution by a factor of 100 or more. Such an optimization produces little overhead on the software side, because the individual characters have to be separated anyway to be recognized.

4.4 Debora Text and Layout

```
MODULE Thin;
    STRUCTURE Algorithm;  (* parallel thinning algorithm for one pixel *)
        INPUT p[8];  (* neighbors of the actual pixel *)
        INPUT even;  (* set of even iterations *)
        OUTPUT del;  (* set if pixel may be deleted *)
        SIGNAL p1[3], p2[3], p3[3], p4[3], p5[3], p6[3], p7[3], p8[3];
        SIGNAL q1[3], q2[3], q3[3], q4[3], q5[3], q6[3], q7[3], q8[3];
        SIGNAL N1[3], N2[3], C[3], a, b, c;
    BEGIN
        p1[0] .= p[0]; p1[1..2] .= 0; q1 .= p1 | p2;  (* algorithm from Guo and *)
        p2[0] .= p[1]; p2[1..2] .= 0; q2 .= p2 | p3;  (* Richard for the deletion *)
        p3[0] .= p[2]; p3[1..2] .= 0; q3 .= p3 | p4;  (* of a pixel depending on *)
        p4[0] .= p[3]; p4[1..2] .= 0; q4 .= p4 | p5;  (* the state of the eight *)
        p5[0] .= p[4]; p5[1..2] .= 0; q5 .= p5 | p6;  (* neighbours *)
        p6[0] .= p[5]; p6[1..2] .= 0; q6 .= p6 | p7;
        p7[0] .= p[6]; p7[1..2] .= 0; q7 .= p7 | p8;
        p8[0] .= p[7]; p8[1..2] .= 0; q8 .= p8 | p1;
        C .= (~p2 & q3) + (~p4 & q5) + (~p6 & q7) + (~p8 & q1);
        a .= (C = 1);
        N1 .= q1 + q3 + q5 + q7;
        N2 .= q2 + q4 + q6 + q8;
        b .= (N1 >= 2) & (N2 >= 2) & ((N1 < 4) | (N2 < 4));
        IF even THEN
            c .= (((q2 | ~p5) & p4) = 0)
        ELSE
            c .= (((q6 | ~p1) & p8) = 0)
        END;
        del .= a & b & c
    END Algorithm;
```

```
STRUCTURE Processor;
    INPUT en-, wr-, ad;  (* processor control signals *)
    INPUT even;  (* set of even iterations *)
    INPUT DataIn[8];
    INOUT DataOut[8];  (* tristate outputs *)
    OUTPUT final;  (* set if no pixel has changed *)
    UNIT Func[2]: Algorithm;
    SIGNAL clk, res[2];
    STATE(clk) InReg0[8], InReg1[8];  (* buffer register *)
    STATE(clk) UReg[11], LReg[11], AReg[11];  (* shift registers *)
    STATE(clk) OutReg[6];  (* output shift register *)
    STATE(clk) Final;  (* final flag *)
BEGIN
    clk .= ~en;  (* use processor signal as clock *)
    InReg1 := InReg0; InReg0 := DataIn;  (* buffer register *)
    IF ad THEN  (* load new data *)
        UReg[3..10] := InReg1;
        LReg[3..10] := InReg0;
        AReg[3..10] := DataIn
    ELSE  (* shift registers by two bits *)
        UReg[3..8] := UReg[5..10];
        LReg[3..8] := LReg[5..10];
        AReg[3..8] := AReg[5..10]
    END;
    UReg[0..2] := UReg[2..4];
    LReg[0..2] := LReg[2..4];
    AReg[0..2] := AReg[2..4];
    Func.even .= even;
    Func.p[0] .= UReg[0..1]; Func.p[1] .= UReg[1..2];  (* connect *)
    Func.p[2] .= UReg[2..3]; Func.p[3] .= AReg[2..3];  (* neighbours *)
    Func.p[4] .= LReg[2..3]; Func.p[5] .= LReg[1..2];
    Func.p[6] .= LReg[0..1]; Func.p[7] .= AReg[0..1];
    res .= AReg[1..2] & ~Func.del;  (* delete pixel if possible *)
    IF res # AReg[1..2] THEN (* update final flag *)
        Final := 0
    ELSE
        Final := Final
    END;
    final .= Final;
    OutReg[0..3] := OutReg[2..5];  (* output shift register *)
    OutReg[4..5] := res;
    IF en & ~wr THEN  (* drive outputs *)
        DataOut[0..5] .= OutReg;
        DataOut[6..7] .= res
    END
END Processor;
END Thin.
```

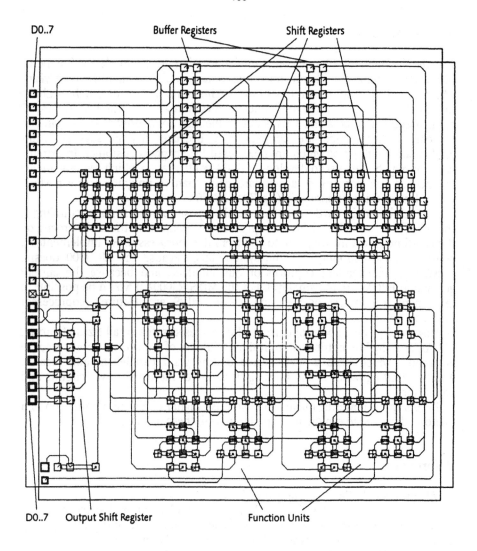

Fig. 4. Layout of Character Thinning Processor

5 Conclusions

The development of Chameleon, from conception to a completely functional prototype running the Oberon operating system, took less than 0.5 man years, which is an important lesson of its own: FPGAs not only provide much flexibility but also keep the complexity of Chameleon low, even when taking the standard IO configurations into account. This is due to the high complexity of some standard IO chips. These chips are complex only because they possess so many configurations, most of which are never used by any given customer. In extreme cases, it may be quicker to develop an FPGA configuration which re-implements the needed functionality of a standard chip, than to program a driver which correctly uses this standard chip (e.g. Zilog Z8530).

A crucial point in the design of a general purpose coprocessor is its coupling to the main processor. The tighter a coprocessor is coupled to the main processor, the less time is lost moving data to and from the coprocessor. We implemented a relatively loose coupling over the system bus, because the LR33000 doesn't provide an independent coprocessor bus interface, and because the current generation FPGAs are too small yet. For the MIPS R3000A it would be possible to implement a coprocessor adhering to the MIPS architecture's coprocessor protocol.

However, the high speed of current RISC processors, the small sizes of current FPGAs, and the bus bottleneck limit the usefulness of a universal coprocessor to algorithms which have a high degree of fine-grain parallelism and locality. Another limiting factor is the missing flexibility of the usual driver circuits and connectors.

FPGAs are a powerful means to customise digital hardware. Chameleon is a platform designed to exploit and experiment with these new facilities.

References

1. Algotronix Ltd.: CAL1024 Datasheet (1990)
2. P. Bertin, D. Roncin, J. Vuillemin: Introduction to Programmable Active Memories. Research Report No. 3, DEC Paris Research Laboratory (1989)
3. K. M. Chandy, J. Misra: Parallel Program Design: A Foundation. Addison Wesley (1988)
4. Z. Guo, W. H. Richard: Parallel Thinning with Two-Subiteration Algorithms. Communications of the ACM, vol. 32, no. 3, pp. 359–373 (1989)
5. B. Heeb: Debora: A System for the Development of Field Programmable Hardware and its Application to a Reconfigurable Computer. Ph. D. thesis no. 10049, ETH Zürich (1993)
6. LSI Logic Corporation: LR33000 MIPS Embedded Processor User's Manual.
7. C. Pfister: CALLAS: A Physical Design Framework for Configurable Array Logic. Ph. D. thesis no. 9940, ISBN 3 7281 1967 9, ETH Zürich (1992)
8. M. Reiser, N. Wirth: Programming in Oberon, Steps Beyond Pascal and Modula. Addison-Wesley (1992)
9. N. Wirth, J. Gutknecht: Project Oberon, The Design of an Operating System and Compiler. Addison-Wesley (1992)

A Highly Parallel FPGA-Based Machine and its Formal Verification

Paul Shaw and George Milne

HardLab
Department of Computer Science
University of Strathclyde
Richmond Street
Glasgow
Scotland, UK

Abstract. The SPACE machine is introduced as a new type of computer architecture, capable of very fast simulation of highly concurrent systems. The machine is designed to be scalable, constructed from a vast array of boards. The decisions made in the the design of the board are discussed, and the actual hardware (based on an array of Field Programmable Gate Array chips) is described. It is shown that this machine can be programmed by translating a subset of the Occam language into asynchronous modules. Using the Circal process algebra, a new method of formally verifying asynchronous modules for these circuits is presented. This method allows bounded gate delays to be included in a two-level modelling mechanism.

1 The SPACE Machine

The SPACE machine (Scalable Parallel Architecture for Concurrency Experiments), a novel architecture based on FPGAs, is a simulation surface where highly concurrent physical phenomena may be emulated, and control-based and systolic systems developed and prototyped.

On the current SPACE board, there exists 9 FPGA chips in the form of Configurable Array Logic (CAL) CAL1024 chips [15], and an Inmos Transputer for administration, memory intensive and more coarse-grained tasks. The board can be connected with others, identical to itself, to form an arbitrarily sized machine. The board is shown in figure 1.

The underlying philosophy for the machine is one of realizing the inherent concurrency of the physical system to be simulated using the fine-grained concurrency of the digital hardware, with a one-to-one mapping between. This clearly differs from conventional von Neumann architectures where the state of the system is encoded and held in memory, with performance restricted by memory access times. On the SPACE machine, the localized state of the physical system is represented by a local area of the machine hardware array.

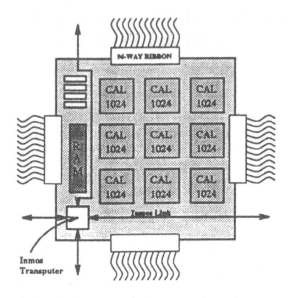

Fig. 1. The SPACE Board

1.1 Design Principles

Although our architecture abandons the notion of a regular matrix of random access memory cells, there is much to be said for organizing an assembly out of regular arrays of components.

Regularity. Regularity economizes on mental effort. A regular assembly of components is easier to design, understand, and to use, than an irregular assembly. A regular assembly is easier to test since a single test procedure has only to be iterated to apply to the whole system.

We have designed our machine according to the principles of regular and modular construction. This regularity can be seen at three levels of scale. Regularity at the chip level, at the board level and of the whole assembly.

The Algotronix CAL chips have a very regular internal structure. They are composed of a square grid of 1024 cells, with North...West bidirectional communication between neighbouring cells. Each cell also contains a logic unit that can implement any two-input boolean logic function or act as a latch. These two inputs are taken from any cell input or from one of two global lines. Each output port may be driven either by an input or by the function unit.

If CAL chips are placed in a grid with their corresponding pins connecting, the result is a grid of cells that appears to stretch seamlessly across the boundary between the chips. As shown in figure 1, the array of chips is 3 × 3 giving 9216 basic computational cells. Two boards with 16 CAL chips each are under construction, giving 16384 cells per board.

The boards are constructed to allow all the data signals from the array of CAL chips to be taken to the edges of the board and brought out onto 96 way ribbon cables. Similarly, the link lines from the Transputer are brought out onto corresponding connectors on the edges of the boards. The consequence is that large arrays of basic cells may be constructed.

2 Compilation

The programming of the SPACE machine would be much simplified if one programming language is used for both the CAL chips and the Transputer. There is a disparity here. Reconfigurable logic technologies are generally programmed via a graphical interface, whereas Occam [14] or C is often used to program a Transputer. How do we bridge this programming gap?

It has been shown [4, 16, 21] that it is possible to translate Occam-like languages directly into asynchronous hardware, thus allowing a meshing of programming paradigms. Asynchronous circuitry has been more intensively studied of late. [5, 9, 10] provide a general study of asynchronous and delay-insensitive circuits, while [11, 12, 20] describe some practical applications.

Asynchronous hardware is especially suited to control-based systems. The natural solution to handling asynchronously occurring events is to use an asynchronous circuit. However, as well as being more natural, performance can also be increased. For example, in a synchronous circuit, events occurring on input ports may have to wait a entire clock cycle to be responded to. In an asynchronous circuit, the response can be immediate.

Regarding cellular automata and systolic problems, synchronous circuitry tends to be smaller and faster. A compiler from Occam to regularly laid out synchronous logic is under development.

2.1 Methodology

Our compilation methodology begins by examining the basic control and data operators of Occam. For each of these, we design a small set of CAL cells with analogous behaviour. Each of these components has a delay-insensitive interface; its correct operation does not depend upon the lengths of its input or output wires. We then use appropriate rules to translate the Occam source into a circuit made up of these delay-insensitive components. Designing these basic components is a necessary step in constructing delay-insensitive circuits, since logic gates prove to be an inadequate starting base [6].

The Occam code currently accepted by the compiler is a subset of full Occam, where the only basic data types are boolean and channels of boolean, with arrays of these types being catered for.

3 Formal Verification of Asynchronous Circuit Modules

In this section, we show how the method by which asynchronous modules used by the Occam compiler can be verified using the Circal [18, 19] process algebra, realized in the form of the Circal-System developed in HardLab.

The Circal-System is a set of software tools built around Circal that provide a framework in which to design, analyse and automatically verify digital hardware. Circal has been shown to be useful in the formal verification of both synchronous [1] and asynchronous circuits [3, 2]. The XCircal [17] hardware description language is used to interface with the Circal-System. This language may be thought of as Circalwith a good deal of syntactic sugar added; XCircal is to Circal what Occam is to CSP [13].

XCircal is an event-based language; processes interact by participating in events and a set of simultaneous events is termed an action. For an action to have occurred, all processes which *may* do the action must participate (a process may do the action if it has the event mentioned within its specification at some point). Primitives of the language are as follows:

- **State Definition:** P <- Q means that process P is defined to have the behaviour of term Q.
- **Termination:** /\ (pronounced 'delta') is the deadlock state, from which the process cannot evolve.
- **Guarding:** x R is a process that will synchronize to perform event x, then behave as R. (a b)S will synchronize with events a and b simultaneously, and then behave as S.
- **Choice:** P + Q is a term that can choose between the actions in process P or those in Q. The choice is dependent upon the environment in which the process is executed.
- **Non-determinism:** P & Q is like choice, except that the choice is made randomly by the processes itself, independent of its environment.
- **Composition:** P*Q 'runs' P and Q in parallel, with synchronization occuring over similarly named events.
- **Abstraction:** P - a hides event set a from P. The actions in a become unobservable.

XCircal also provides a Bool data type, useful for modelling the states of wires. This is composed of two events, for Bool x, there are associated events x.0 and x.1.

We shall use these features of XCircal in the following pages to describe and verify an asynchronous design methodology suitable for the SPACE machine.

3.1 Traditional Bounded Gate-Delay Models

The asynchronous circuits models used here are *bounded delay* models, as described in [7, 8], where gates (or any combinatorial logic circuit) have an associated bounded delay. The reason for the introduction of these models is to allow non-deterministic physical properties of devices to be absorbed in an abstract model. Wires can also be given bounded delays by considering them to be gates with no logical function.

The model described in [7, 8] dictates that gates must operate by three logic levels rather than two. An 'uncertain value' is added to the traditional

HIGH/LOW scheme, where an 'X' signifies the uncertain value. Gate models are adjusted such that the uncertain value is incorporated. For instance, for an AND gate, the output is definite when either input is 0 or both inputs are 1. The output is uncertain in all other cases.

To create a gate with bounded delay, an bounded delay 'box' is connected to the output of a delayless gate. This box has associated with it lower and upper time bounds (written $[l,u]$), within which the exact (physical) delay will definitely lie. When this box detects a change to a 0 or 1 on its input, it counts up to the lower bound. When reached, it outputs the uncertain value X until the upper delay bound is reached. Then, the 0 or 1 is output. When the bounded delay box's input changes to an X, this change is echoed to the output when the lower bound time is reached.

The input to the bounded delay box must remain constant at either 0 or 1 for the upper bound time. Anything shorter than this is considered a spike, highlighting incorrect circuit operation.

This model has the effect of *narrowing* the window where a signal is retained at a value: uncertainty values are inserted instead. This means that in feedback loops, transitions must be a reasonable distance apart. If not, the circuit will generate a spike condition or degenerate to emitting an infinite stream of uncertain values. Through these criteria, race conditions in circuits can be detected. Violations of setup and hold times are just special cases of such races.

3.2 XCircal and the Three-Valued System

The methods described in the previous section could be applied directly to XCircal. Normally, modelling is done in XCircal using two events per wire W to describe its current state (*e.g.* W.0 and W.1). However, this could easily be extended to include the uncertain value. Each wire could modelled by three events, W.0, W.X, and W.1. Delayless gates and bounded delay boxes would be based around this three-valued system.

This however, is not a desirable option. It makes little sense to use three-level models when much theory has been devoted to (and a great many models constructed from) two-level models. New three-level models would be incompatible with old ones, and the experience gained in the development of two-level models would be redundant.

For this reason, creation of a bounded delay model using only two-level logic is highly desirable. In the next section, we will shown how this can be achieved in a simple and elegant manner.

3.3 The XCircal Solution

The non-deterministic choice operator is the way to describe uncertainty in the XCircal hardware description language; P & Q means that the process will randomly engage in process P or Q. The choice is made by the process itself, not by its environment. Non-deterministic choice can be used together with a two-level scheme to model bounded delays.

The way a bounded delay box is modelled is as follows: First, all edges are delayed until the lower bound is met. After this point, and at each discrete time unit until the edge is issued at the output, the delay box *non-deterministically* decides whether to issue the edge now, or wait. If the upper time bound is reached, and the edge has still not been issued, it is done at that point. An XCircal model of a delay box with bounds [1,2] is shown below. Delay boxes with different bounds can be created and have the same basic structure.

```
Process BoundedDelay1_2(Bool in,out)
{
Process DEL0,DEL1,DEL10,DEL01

   DEL0  <- (in.0 out.0 t)DEL0 + (in.1 out.0 t)(DEL1 & DEL10)
   DEL1  <- (in.1 out.1 t)DEL1 + (in.0 out.1 t)(DEL0 & DEL01)
   DEL10 <- (in.1 out.0 t)DEL1 + (in.0 out.0 t)/\
   DEL01 <- (in.0 out.1 t)DEL0 + (in.1 out.1 t)/\
   return(DEL0)
}
```

The method of modelling discrete time is that all actions synchronize with a global t event; t is assumed to occur at regular intervals. The delay box is prepared in starting state DEL0 (by returning this state), corresponding to the steady state with the output low. The input and output of the delay box are level-based; the 0 and 1 extensions describe the *level* of the wire at that port. When two edges are propagating through the delay box at any one time, this is considered erroneous, and the process deadlocks. The importance of this deadlock will be discussed later.

3.4 Target Medium: The CAL Surface

Construction of bounded delay models is the essence of the modelling mechanism described above. Describing differing technologies is simply a matter of using different bounded delay elements. On the current chip (the CAL1024), computation of a function takes between 6ns and 8ns while routing through a cell takes 2ns. Time can therefore be quantized to 2ns intervals. Gates then have bounded delays of [3,4] , and routing delay 1. A CAL gate model is built by composing a delayless model of the desired gate with a [3,4] delay box. The delayless gate can be specified as follows:

```
Process IdealAnd(Bool a,b,c)
{
Process AND

   AND <- (a.0 b.0 c.0)AND + (a.0 b.1 c.0)AND +
          (a.1 b.0 c.0)AND + (a.1 b.1 c.1)AND
   return(AND)
}
```

and the composition as:

```
Process CALAnd(Bool A,B,C)
{
Bool s

    return(IdealAnd(A,B,s)*BoundedDelay3_4(s,C) - s)
}
```

The composition is shown graphically in figure 2.

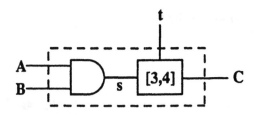

Fig. 2. Model of a CAL AND Gate

Asynchronous operators made from these gate models can be built by further composition. Figure 3 shows a 4 CAL cell layout of a Muller C element, and, for clarity, an equivalent schematic. The C element performs a synchronization function. When its inputs are equal, the state of the inputs is copied to the output, otherwise the output remains as it is. Both inputs to the C element must change to receive a subsequent output change. In figure 3, all gates have delay, and wires have zero delay except those marked with a 1, which have a unit delay. Such delays correspond to wires that connect *through* a CAL cell, and as such have a 2ns delay associated.

The C element can now be described in XCircal as:

```
Process CElement(Bool A,B,C)
{
Bool Adelayed,Bdelayed,AandB,AorB,loop

    return(CALAnd(B,Adelayed,AandB)*CALOr(A,Bdelayed,AorB)*
           BoundedDelay1(A,Adelayed)*BoundedDelay1(B,Bdelayed)*
           CALAnd(AorB,loop,C)*CALOr(AandB,C,loop) -
           (Adelayed Bdelayed AandB AorB loop t))
}
```

This composition results in a process describing the possible actions of this implementation of a C element. Notice that time (global event t) has been abstracted away. This means that no absolute time difference now exists between any two events occurring at the C element's ports. Only information about the time-ordering of events remains.

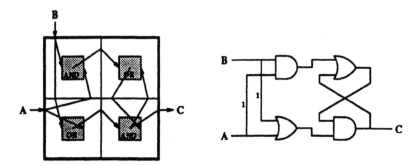

Fig. 3. CAL and Schematic Muller C Elements

3.5 Converting to Edges

Specifications of asynchronous components are generally *edge-based* while model of the implementation of a C element is level-based. As we require to perform a verification between the two, some form of conversion must occur. For this, we use a level-to-edge converter placed on each port of the C element. The converter is specified in XCircal as:

```
Process Lev2Edge(l,e)
{
Process LOW,HIGH

   LOW  <- (1.0)LOW  + (1.1 e.1)HIGH
   HIGH <- (1.1)HIGH + (1.0 e.0)LOW
   return(LOW)
}
```

Since XCircal has no concept of input or output, we can use the same edge converter for all ports of the C element. It should perhaps be mentioned at this point that all XCircal models have been initialized in a state such that their ports are in the low state. This is done for clarity: models are kept simpler, are easier to understand, and suffice for the purposes of this paper. An edge-based C element can be described in XCircal:

```
Process EdgeCElement(Bool A,B,C)
{
Bool x,y,z

   return(CElement(x,y,z)*Lev2Edge(x,A)*
          Lev2Edge(y,B)*Lev2Edge(z,C) - (x y z))
}
```

3.6 Verification

Verification of correctness of our C element requires two further pieces of information to be provided. The first of these is the specification of the C element. The second is a description of the *operating constraints* of the circuit. This is essentially a simulated environment for the C element. It must provide inputs at the correct time, and malfunction if the C element delivers spurious output.

The edge-based specification of the C element can be described in XCircal as:

```
Process CElementSpec(Bool A,B,C)
{
Process CLOW1,CLOW2,CHIGH1,CHIGH2

   CLOW1  <- (A.1)(B.1)CLOW2 + (B.1)(A.1)CLOW2 + (A.1 B.1)CLOW2
   CLOW2  <- (C.1)CHIGH1
   CHIGH  <- (A.0)(B.0)CHIGH2 + (B.0)(A.0)CHIGH2 + (A.0 B.0)CHIGH2
   CHIGH2 <- (C.0)CLOW1
   return(CLOW1)
}
```

Here, 0 and 1 extensions denote either a negative or positive *edge* on the named port.

The next step to to define an operating environment for the C element. This environment must be able to detect erroneous edges issued from the C element's output port, as well as driving it correctly on its input ports. A process to achieve this is described as follows:

```
Process CElementEnv(Bool A,B,C)
{
Process M000,M001,M010,M011,M100,M101,M110,M111

  M000 <- (A.1)M100 + (B.1)M010 + (A.1 B.1)M110 + (C.1)/\
  M001 <- (C.0)M000
  M010 <- (A.1)M110 + (C.1)/\
  M011 <- (B.0)M001 + (C.0)/\
  M100 <- (B.1)M110 + (C.1)/\
  M101 <- (A.0)M001 + (C.0)/\
  M110 <- (C.1)M111
  M111 <- (A.0)M011 + (B.0)M101 + (A.0 B.0)M001 + (C.0)/\
  return(M000)
}
```

When the environment process detects an edge from the output of the C element occurring at the wrong point of time, it deadlocks.

We can now attempt to verify the correctness of the C element, by executing the following fragment of XCircal:

```
Bool in1,in2,out

print ("(CElementSpec(in1,in2,out)) ==
       "(EdgeCElement(in1,in2,out)*CElementEnv(in1,in2,out)))
```

The Circal-System indicates equivalence by printing **true** or **false** (this particular verification evaluates to **true**). The ˜ is an operator that expands out the process description to one which can be formally verified, and the **==** operator denotes testing equivalence. In particular, non-deterministic processes (our delay boxes) *must* pass the test (*i.e.* all possible paths of the non-deterministic process are traced to ensure that *none* produce actions contrary to the specification). That the processes *may* pass the test because of a lucky sequence of non-deterministic actions is insufficient for our purposes.

It now becomes more obvious why we earlier moved processes to a dead-locked state when we discovered that the circuit was operating erroneously. If any process deadlocks owing to the incorrect operation of its environment, the entire system will become deadlocked. No deadlock exists in the *specification* of the circuit, and as such the verification will fail.

Not all verifications are as straightforward as the one described here and quite often the verification fails. We have found that the common problem here is that the output changes before the internal feedback loop has had time to stabilize. When the environment detects this change, it may change the input immediately. Since the feedback loop has not stabilized, either deadlock occurs (as detected by a delay box) or a spike appears at the output contrary to specification. This type of failure is solved by placing a sufficiently large delay at the output(s) of the circuit, such that the environment sees the output(s) change later.

3.7 Discussion

A methodology has been introduced for the verification of asynchronous circuits. If this verification effort was coupled with a proof of correctness of the Occam compiler, it would provide us with a very high level of confidence that *all* circuits generated by the compiler are correct.

The bounded gate delay model (and consequently circuit models) introduced operate by two-levels as opposed to the traditional three. We believe that this new model is the correct and natural form to use in the context of XCircal, given its powerful non-determinism operator and corresponding verification tools. A simplicity and elegance has been achieved by staying with two-level logic which would have been corrupted by moving to three.

4 Conclusion

A new scalable parallel computer based on FPGAs has been introduced. The implementation mixes both traditional computational components and FPGA logic while using parts that are completely scalable, using a local four-neighbour grid connectivity pattern. Throughout the design, the concepts of modularity and regularity have been adhered to so as to provide a simple and efficient framework.

It has been shown that this machine can be programmed in a subset of Occam which (as well as being able to be executed by the Transputer) can be compiled directly into a digital hardware configuration file suitable for downloading to the CAL chips.

Finally, we have presented a novel method of verifying asynchronous circuits using a bounded-delay model, but only utilizing two-level logic. These verifications are done automatically using the Circal-System , and can be simply adapted to technologies other than Algotronix CAL chips. Coupled with a proof of correctness of the compilation algorithm, we would be provided with a very high level of confidence in the correctness of *all* circuits produced by the compiler.

The SPACE machine is one possible future of a computing machine which VLSI technology has been able to create. The power of reprogrammability of such a machine changes the concept of a general purpose computer to one which is radically different from the traditional von Neumann machine.

References

1. Andrew Bailey, George McCaskill, Jim McIntosh, and George Milne. The description and automatic verification of digital circuits in Circal. In P. Camurati and P. Prinetto, editors, *Proceeding of the Advanced Research Workshop on Correct Hardware Design Methodologies*. Elsevier/North-Holland, 1992. Workshop held from June 12–14, 1991, Turin, Italy.

2. Andrew Bailey, George A. McCaskill, and George J. Milne. An exercise in the automatic verification of asynchronous designs. Technical Report HDV–26–93, Department of Computer Science, University of Strathclyde, January 1993.

3. Andrew Bailey and George Milne. Using Circal to analyse Sutherland's asynchronous micropipeline design style. Technical Report HDV–13–91, Department of Computer Science, University of Strathclyde, May 1991.

4. Erik Brunvand and Robert F. Sproull. Translating concurrent communicating programs into delay-insensitive circuits. Technical Report CMU–CS–89–126, School of Computer Science, Carnegie Mellon University, April 1989.

5. J. A. Brzozowski and J. C. Ebergen. Recent developments in the design of asynchronous circuits. Technical Report CS–89–18, Computing Science Department, University of Waterloo, May 1989.

6. J. A. Brzozowski and Jo C. Ebergen. On the delay-sensitivity of gate networks. Computing Science Notes 90/05. Department of Mathematics and Computing Science, Eindhoven University of Technology, June 1990.

7. J. A. Brzozowski and C-J. Seger. A unified framework for race analysis of asynchronous networks. *Journal of the Association of Computing Machinery*, 36:20–45, 1989.

8. J. A. Brzozowski and Michael Yoeli. On a ternary model of gate networks. *IEEE Transactions on Computers*, 28:178–184, 1979.

9. Jo C. Ebergen. A formal approach to designing delay-insensitive circuits. *Distributed Computing*, 5:107–119, 1991.

10. Jo C. Ebergen. Parallel computations and delay-insensitive circuits. Technical Report CS–91–05, Department of Computer Science, University of Waterloo, January 1991.

11. Jo C. Ebergen and Sylvain Gingras. An asynchronous stack with constant response time. Submitted to the TALI '92 workshop, 1992.

12. Jo C. Ebergen and Ad M. G. Peeters. Modulo-n counters: Design and analysis of delay-insensitive circuits. In J. Staunstrup and R. Sharp, editors, *Proceedings of the Second IFIP Workshop on Designing Correct Circuits*. North-Holland, 1992.

13. C. A. R. Hoare. *Communicating Sequential Processes*. Prentice-Hall, 1985.

14. Inmos. *A Tutorial Introduction to* occam *Programming*. BSP Professional Books, 1989.

15. Tom Kean. *Configurable Logic: A Dynamically Programmable Cellular Architecture and its VLSI Implementation*. PhD thesis, University of Edinburgh, Dec 1989.

16. Alain J. Martin. Compiling communicating processes into delay-insensitive VLSI circuits. *Distributed Computing*, 1:226–234, 1986.

17. George McCaskill. The XCircal user guide and reference manual. Technical Report HDV–18–91, Department of Computer Science, University of Strathclyde, October 1991.

18. George J. Milne. Circal and the representation of communication, concurrency and time. *Association of Computing Machinery Transactions on Programming Languages and Systems*, 7(2), 1985.

19. George J. Milne. The formal description and verification of hardware timing. *IEEE Transactions on Computers*, 40(7), 1991.

20. Ivan. E. Sutherland. Micropipelines. *Communications of the Association of Computing Machinery*, 32(6):720–738, June 1989.

21. Kees van Berkel, Joep Kessels, Marly Roncken, Ronald Saeijs, and Frits Schalij. The VLSI-programming language Tangram and its translation into handshake circuits. In *The European Design and Automation Conference*, 1991.

FPGA based
Self-Test with Deterministic Test Patterns

Arno Kunzmann

Forschungszentrum Informatik (FZI)

Haid-und-Neu-Strasse 10-14

7500 Karlsruhe 1

Germany

Phone +49 721 9654 456

Fax   +49 721 9654 409

Abstract. This paper describes a new approach to synthesize a cost-efficient self-test hardware for a given set of deterministic test patterns. To minimize the test hardware effort, instead of all the patterns only a very small subset has to be selected such that an easy generation of all necessary test patterns is ensured. This procedure drastically decreases the storage requirements (over 80%) and therefore reduces distinctly the self-test hardware effort. The realization of an external self-test by a specific test chip was done with XILINX FPGAs, since field-programmable gate-arrays are best-suited for applications with a low production volume. Experimental results on all the ISCAS benchmark circuits underline the efficiency of our approach.

1 Introduction

In the field of VLSI test there exist several reasons for a minimization of test sets. The main objectives are a shortening of the test application time especially for large scan-based circuits and a reduction of the storage requirements. The latter is of special interest for this paper since a specific test chip for deterministic test pattern generation will be proposed and realized. This solution can be used for an external self-test, but the approach can also be adapted for an internal self-test. Since up to now the generation of deterministic test patterns by different types of (non-) linear shift registers has not been very promising and requires much hardware (e.g. [4]), conventional RAM-based approaches are still of very importance. Obviously the test hardware effort is dominated by the RAM size.

In this paper a new approach will be presented that significantly reduces the storage requirements since only a small subset of a given deterministic test pattern set has to be stored. Based on this subset all applicable test patterns can easily be derived and realized with very low effort by a specific test chip. The realisation of the test chips with FPGAs will underline the efficiency of this proposal.

2 Related Work

In order to discuss the important work in the field of test set minimization two different methods can be distinguished: (1) the test pattern generation algorithm itself is modified and (2) the minimization procedure is independent of the test generation algorithm and can be used in an additional post-processing phase.

The proposals for a modification and extension of test generation algorithms are mainly based on an analysis of the fault sets [2]. This includes the dynamic test compaction [5] where tests for additional faults are generated as long as sufficient unspecified inputs are available. In [8] an innovative approach is presented to further reduce the number of test patterns. The basic idea of this procedure consists in unspecifying those primary inputs that are not essential for the fault detection. By this procedure the average number of faults detected by a test pattern can be enlarged. Experimental results, based on the PODEM algorithm, showed the effectiveness of this improved dynamic compaction. In [9] a procedure for the generation of independent fault sets is presented. Tuned on these fault sets an efficient test set reduction can be performed, but this approach is restricted to small and medium sized circuits.

In a post-processing phase after the completion of the test pattern generation static compactions can be performed [1], typically by a resimulation of the test patterns in reverse order [10]. But the computation of a minimal test pattern set is also a NP-complete problem that in general can only be solved by heuristics.

In this paper we propose a new procedure which combines the two ideas of minimizing the number of deterministic test patterns to be stored and enabling the synthesis of an efficient self test hardware structure. Furthermore this procedure can be applied in a post-processing phase and is based on a set of deterministic test patterns. Therefore no modifications or extensions of the test pattern generation algorithm are required.

3 Selection of Minimized Test Sets

The basic idea of our approach can be explained by a simple observation if deterministic test patterns are compared: very often pairs of test patterns can be found that differ in only one bit position. Starting with stuck-at faults at primary inputs the sensitized paths for fault propagation remain unchanged, only the corresponding bit position has to be inverted. Similar observations can also be made for internal faults for given fault propagation paths. Of course it is not ensured than only one bit position has to be changed to detect additional faults, but we will show later that this is often the case. A further important observation is the sensitizing of operation units. Suppose for instance a multiplication unit: here by any input combination, paths are sensitized and faults will be propagated.

These observations led to the approach to derive from each deterministic test pattern all distinct patterns that differ only in a single bit from the original. Given a test pattern of width w, w additional test patterns can be derived. This pattern set is the correspondent "toggle pattern set".

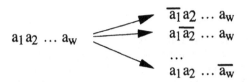

Fig. 1 Test pattern and its toggle pattern set

To give a first example of the efficiency of this approach we analyze the combinational benchmark circuit c6288. To obtain full fault coverage 35 test patterns are generated with the SPROUT9v algorithm [6]. To maintain full coverage it is sufficient to select only 7 test patterns of width 32. Out of this basic set 224 additional test patterns have to be derived for test application. Obviously the storage requirement is significantly reduced, at the same time offering an easy generation of the overall test pattern set by test hardware (see next section).

Now the question will be discussed how for a given set of deterministic test patterns a minimized basic test set can be derived. The basic selection procedure consists of the following two phases:

(1)   Computation and minimization of a basic test pattern set
(2)   Computation of the bit positions to be inverted

Each task can be solved by an additional fault simulation phase. The main difference is the order of the test patterns. For phase (1) after each deterministic test pattern all the derivable test patterns are grouped and fault simulated. Only if a group detects at least one additional fault the corresponding original deterministic test pattern will be part of the basic test pattern set. In [7] an additional minimization procedure is described in detail.

Based on this basic test pattern set the question has to be answered whether it is necessary to invert each bit position of the basic test patterns. This answer will be given by phase (2) as a direct result of a second fault simulation with a different grouping of test patterns (w denotes the pattern width):

1.   All basic test patterns
2.   All basic test patterns with inverted position 1
3.   All basic test patterns with inverted position 2
...   ...
$w+1$.   All basic test patterns with inverted position w

Bit position j has to be inverted iff group $j+1$ detects at least one additional fault that was not yet detected by one of the previous groups.

4 Synthesis of the Test Chip

The proposed selection strategy is well suited for self test and/or external test generation by special test chips. Figure 2 shows the principal structure of the test generation hardware.

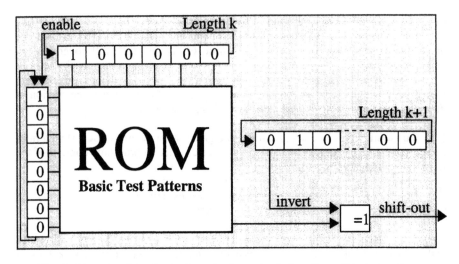

Fig. 2 Structure of synthesized test hardware to derive all possible test patterns

The most complex module of the test chip is the RAM containing all deterministic basic test patterns. For the row-selection a simple shift register can be used with the shown initialization. The selection of the bit position is enabled by a cyclic shift register, also initialized by all zeros except a single one. The inversion of selected bit positions can be performed by a third register of length k+1, k denoting the pattern width. Starting with the above shown initialization, the selected test pattern will be shifted out without any change. During the following k cycles always one (different) bit position will be inverted and sent to the chip to be tested.

If only some selected bit positions have to be inverted the shown structure has to be modified: instead of a shift register a counter module can be inserted as depicted in figure 3.

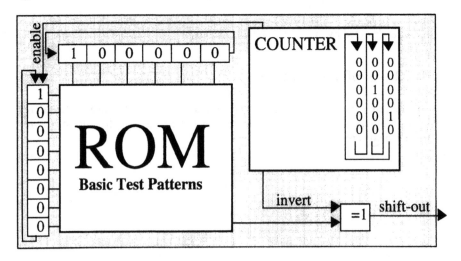

Fig. 3 Structure of synthesized test hardware to derive selected test patterns

Only if the counter reaches certain states the "invert" signal is set to 1, otherwise held at 0. Furthermore the counter enables the shift register for row-selection to move on to the next test patterns.

To compare the hardware effort of the proposed procedure with a conventional test generation, a realization as field-programmable gate-arrays will be analyzed in the following section. To include the option to invert selected bit positions the comparison will be based on the structure of figure 3.

5 Experimental Results

The described procedure was applied to all the ISCAS-85 and ISCAS-89 benchmark circuits [3,12] and compared with the minimization algorithms of [9] and [8]. Table 1 gives an overview on the original, the minimized test pattern sets and the size of the basic test pattern set. For the seven largest sequential circuits a complete scan design is assumed.

Circuit	Original Number of Determ. Patterns	Determ. patterns [9]	Determ. Patterns [8]	Minimized Number of Basic Test patterns	Relative Reduction Compared with Original Number
c432	67	30	n.a.	16	76.1%
c499	62	52	n.a.	10	83.9%
c880	67	21	30	14	79.1%
c1355	96	82	86	16	83.3%
c1908	136	n.a.	126	20	85.3%
c2670	127	(30)	67	20	84.3%
c3540	197	n.a.	111	35	82.2%
c5315	168	n.a.	56	25	85.1%
c6288	35	n.a.	16	7	80.0%
c7552	227	n.a.	87	38	83.3%
s5378	315	n.a.	111	62	80.3%
s9234	481	n.a.	149	78	83.8%
s13207	578	n.a.	237	91	84.3%
s15850	500	n.a.	123	76	84.8%
s35932	74	n.a.	13	14	81.1%
s38417	1107	n.a.	95	166	85.0%
s38584	883	n.a.	125	146	83.5%
Total	5120		1432	834	83.7%

n.a.: not available   ( ) : value for irredundant circuit

Table 1 Comparison of test compaction methods

Compared with the original test pattern number the average relative saving is about 83%. Table 1 also shows that these savings are rather independent of the circuit size and of the test pattern width.

A direct comparison with [8] results in 1432 test patterns vs. 834 basic patterns that is a 43.4% reduction. This number underlines the effectiveness of the proposed procedure. But it can be observed that with growing test pattern width the savings are reduced and for the last three benchmark circuits even more test patterns are required. This is caused by the test pattern generation algorithm SPROUT9v: with each test pattern all unspecified inputs are randomly set to 0 or 1 and there is no attempt made to extend a test pattern to cover additional faults as e.g. proposed in FAN [11]. Therefore only a few fault propagating paths are sensitized and the inversion strategy is less effective.

Table 2 shows the resulting test application time in terms of the overall number of test patterns.

Circuit	Min. Number of Basic Test Patterns	Test Pattern Width	Bits to be Inverted	Percentage of Bits to be Inverted	Overall Number of Test Patterns
c432	16	36	27	75.0%	448
c499	10	41	28	68.3%	290
c880	14	60	34	56.7%	490
c1355	16	41	36	87.8%	592
c1908	20	33	33	100.0%	680
c2670	20	233	96	41.2%	1.940
c3540	35	50	38	76.0%	1.365
c5315	25	178	112	62.9%	2.825
c6288	7	32	22	68.8%	161
c7552	38	207	111	53.6%	4.256
s5378	62	215	97	45.1%	6.076
s9234	78	248	146	58.9%	11.466
s13207	91	701	200	28.5%	18.291
s15850	76	612	259	42.3%	19.760
s35932	14	1.764	168	9.5%	2.366
s38417	166	1.665	536	32.2%	89.142
s38584	146	1.465	474	32.4%	69.350

Table 2 Overall number of test patterns

To reduce the number of test patterns for the test application phase only part of the bit positions must be inverted. As depicted in column 5 the average is about 50%. These results also show that in principle this percentage is decreasing with increasing pattern width. The last column shows the total number of test patterns, the product of column 2 and column 4 incremented by 1.

The final table 3 and 4 compare the resulting hardware effort for a test chip, based on the original deterministic test patterns and on the proposed basic test pattern set. For this comparison a realization with the field-programmable gate-arrays (XILINX) is assumed, since this technology is best-suited for application specific circuits with a low production volume.

To compute the necessary number of combinational logic blocks (CLBs), let TPC denote the Test Pattern Count and TPW the Test Pattern Width. The size of the RAM was estimated by

(1) (TPC * TPW) / 32 * 1.3.

Each CLB has a capacity of up to 32 bit, the factor 1.3 was added to include some additional logic for address-encoding. The CLB count of the two shift registers is computed by

(2a) TPC / 2 and

(2b) TPW / 2,

because two flipflops can be realized within a single CLB. Without further optimization the size of the counter can be estimated by

(3) ln(TPW*(BPI+1)) / 2,

BPI denoting the number of bit positions to be inverted. The following two tables show the rounded numbers for all the analyzed benchmark circuits.

Circuit	RAM Size	Address Decoding	Total Number of CLBs
c432	98	52	150
c499	103	52	155
c880	163	64	227
c1355	160	69	229
c1908	182	85	267
c2670	1.202	180	1.382
c3540	400	124	524
c5315	1.215	173	1.388
c6288	46	34	80
c7552	1.909	217	2.126
s5378	2.751	265	3.016
s9234	4.846	365	5.211
s13207	16.460	640	17.100
s15850	12.431	556	12.987
s35932	5.303	919	6.222
s38417	74.878	1386	76.264
s38584	52.552	1174	53.726

Table 3 CLB count based on original test patterns

Circuit	RAM Size	Address Decoding	Counter	Total Number of CLBs	Relative Reduction
c432	23	26	10	59	60.5%
c499	17	26	10	53	65.7%
c880	34	37	11	82	63.8%
c1355	27	29	11	67	70.7%
c1908	27	27	10	64	76.0%
c2670	189	127	14	330	76.1%
c3540	71	43	11	125	76.1%
c5315	181	102	14	297	78.6%
c6288	9	20	10	39	50.9%
c7552	320	123	15	458	78.5%
s5378	542	139	14	695	77.0%
s9234	786	163	15	964	81.5%
s13207	2.592	396	17	3.005	82.4%
s15850	1.890	344	17	2.251	82.7%
s35932	1.003	889	18	1.910	69.3%
s38417	11.025	914	11	11.950	84.3%
s38584	9.046	809	11	9.866	81.6%

Table 4 Comparison of CLB count based on basic test patterns

As expected the relative reduction is in the same range as computed for the deterministic test patterns. With respect to the above depicted numbers of test patterns and an estimated maximum clock frequency of 4 MHz the test application time can be easily kept below one second.

Summary and Future Work

Based on a new method for deterministic test pattern generation by hardware, its realisation by FPGAs was discussed. The results with the ISCAS benchmark circuits proofed the efficiency of this approach. Compared with random test pattern sets the resulting number of test patterns is quite low.

Further research will discuss the combination of deterministic and random test: with an additional linear feedback shift register it would be possible first to generate a certain number of random test patterns and then, for the hopefully few remaining faults, deterministic test patterns are generated. This combination must take into account both the maximum test application time, represented by a maximum number of test patterns with respect to the test pattern width, and the resulting test hardware effort.

Additional research activities will be directed towards an extension of the presented approach to circuits with partial scan design. Here the test system INSPIRATION [6] can be the basis for a planned extension to include also the generation of deterministic test pattern sequences instead of test pattern sets.

Acknowledgement

The author would like to thank Ulrich Weinmann for his valuable contribution to this work, especially for his analysis of the FGPA realisations. I also want to express my appreciation to André Hergenhan for his implementation work of a first synthesis system that generates xnf-descriptions of the resulting self-test chip.

References

1. M. Abramovic et al.: Digital Systems Testing and Testable Designs, Computer Science Press, 1990

2. S.B. Akers et al.: On the Role of Independent Fault Sets in the Generation of Minimal Test Sets, Proc. International Test Conference 1987, pp. 1100 - 1107

3. F. Brglez, H. Fujiwara: A Neutral Netlist of Combinational Benchmark Designs, International Symposium on Circuits and Systems, 1985

4. W. Daehn: Deterministische Testmustergeneratoren für den Selbsttest von integrierten Schaltungen, Ph.D. Thesis, University of Hannover, 1983

5. P. Goel, B.C. Rosales: Test Generation and Dynamic Compaction of Tests, Test Conference 1979, Digest of Papers, pp. 189 - 192

6. A. Kunzmann and H.-J. Wunderlich: An Analytical Approach to the Partial Scan Design, JETTA, Vol. 1, Kluwer, Boston 1990, pp. 163 - 174

7. A. Kunzmann: Generation of Deterministic Test Patterns by Minimal Basic Test Sets, EuroDAC '92, pp. 312 - 317

8. I. Pomeranz, L.N. Reddy, S.M. Reddy: COMPACTEST: A Method to Generate Compact Test Sets for Combinational Circuits, Proc. International Test Conference 1991, pp. 194 - 203

9. G. Tromp: Minimal Test Sets for Combinational Circuits, Proc. International Test Conference 1991, pp. 204 - 209

10. J.A. Waicukauski et al.: ATPG for ULTRA-Large Structured Designs, Proc. International Test Conference 1990, pp. 44 - 51

11. H. Fujiwara: FAN: A Fanout-Oriented Test Pattern Generation Algorithm, International Symposium on Circuits and Systems, 1985, pp. 671 - 674

12. F. Brglez, D. Bryan, K. Kozminski: "Combinational Profiles of Sequential Benchmark Circuits", Int. Symposium on Circuits and Systems, 1989, pp. 1929 - 1934

FPGA Implementation of Systolic Sequence Alignment

Dzung T. Hoang[*][1] and Daniel P. Lopresti[**][2]

[1] Department of Computer Science, Brown University
Providence, RI 02912-1910, USA
[2] Matsushita Information Technology Laboratory
182 Nassau Street, Princeton, NJ 08542-7072, USA

Abstract. This paper describes an implementation of a novel systolic array for sequence alignment on the SPLASH reconfigurable logic array. The systolic array operates in two phases. In the first phase, a sequence comparison array due to Lopresti [1] is used to compute a matrix of distances which is stored in local RAM. In the second phase, the stored distances are used by the alignment array to produce a binary encoding of the sequence alignment. Preliminary benchmarks show that the SPLASH implementation performs several orders of magnitude faster than implementation on supercomputers.

1 Introduction

The work presented in this paper was begun during the first author's summer internship at the National Cancer Institute's Laboratory of Mathematical Biology in Fredrick, Maryland. The goal was to develop genetic sequence analysis algorithms for the SPLASH reconfigurable logic array [2]. A systolic sequence comparison algorithm that computes the edit distance between a pair of sequences had already been implemented on SPLASH [3]. Certain applications of interest to biologists at the laboratory, such as multiple alignment of genetic (DNA and RNA) sequences, however, require more than just the edit distance: a more informative analysis of the similarity, or homology, of the sequences in the form of an alignment is required. In this paper, we describe an implementation of a systolic algorithm for computing sequence alignments on SPLASH. Prior to our work, we know of no systolic array for computing sequence alignments.

1.1 Sequence Comparison and Alignment

Given a source sequence $S = s_1 s_2 \cdots s_m$ and a target sequence $T = t_1 t_2 \cdots t_n$, the *edit distance* between S and T is defined to be the minimum cost of transforming

---

[*] Supported during Summer 1991 by an NIH Summer Internship and afterwards by an NSF Graduate Fellowship.
[**] Supported by NSF grant MIP-9020570.

S to T through a series of the following *edit operations*: deleting a character, inserting a character, and substituting one character for another[3].

In certain applications, such as approximate multiple sequence comparison [4] and protein folding [5], in addition to the edit distance, we need to know the series of edit operations that leads to a minimum cost transformation. A standard way to represent the transformation is with an *alignment*. In an alignment, the characters of the source and target sequences are arranged in a matrix with two rows. The source sequence, possibly with embedded null characters, '−', is placed in the first row. Similarly, the characters of the target sequence are placed in the second row. The matrix is analyzed column-wise. A column containing $\left[\begin{smallmatrix} x \\ - \end{smallmatrix}\right]$ indicates deletion of the character x; a column containing $\left[\begin{smallmatrix} - \\ y \end{smallmatrix}\right]$ indicates insertion of the character y; and a column $\left[\begin{smallmatrix} x \\ y \end{smallmatrix}\right]$ indicates substitution of y for x. A column consisting of two nulls is not allowed. Here is an example of an alignment: $\left[\begin{smallmatrix} A & G & A & C & T & A & - & G & G \\ T & G & - & C & T & A & A & G & C \end{smallmatrix}\right]$. For a given cost function, there may be more than one minimum-cost alignment. The alignment algorithm presented here computes one such alignment.

1.2 Dynamic Programming

The edit distance can be computed sequentially with a well-known dynamic programming algorithm [6,7] in $O(mn)$ time. Let $S = s_1 s_2 s_3 \cdots s_m$ be the source sequence, $T = t_1 t_2 t_3 \cdots t_n$ be the target sequence, and $d_{i,j}$ be the edit distance between the subsequences $s_1 s_2 \cdots s_i$ and $t_1 t_2 \cdots t_j$. Then

$$d_{0,0} = 0 \ ,$$
$$d_{i,0} = d_{i-1,0} + c_{\mathrm{del}(s_i)}, \ 1 \le i \le m \ ,$$
$$d_{0,j} = d_{0,j-1} + c_{\mathrm{ins}(t_j)}, \ 1 \le j \le n \ ,$$

and

$$d_{i,j} = \min_{\substack{1 \le i \le m, \\ 1 \le j \le n}} \begin{cases} d_{i-1,j} + c_{\mathrm{del}(s_i)} \ , \\ d_{i,j-1} + c_{\mathrm{ins}(t_j)} \ , \\ d_{i-1,j-1} + c_{\mathrm{sub}(s_i,t_j)} \ . \end{cases}$$

Here $c_{\mathrm{del}(s_i)}$ is the cost of deleting s_i, $c_{\mathrm{ins}(t_j)}$ is the cost of inserting t_j, and $c_{\mathrm{sub}(s_i,t_j)}$ is the cost of substituting t_j for s_i.

An alignment can be constructed by creating pointers to indicate the minimization choices when evaluating the dynamic programming recurrence. An example dynamic programming table augmented with pointers is shown in Fig. 1. By tracing a path from the lower-right corner to the upper-left corner, we can construct an alignment in reverse. The bold pointers in Fig. 1 show the path that corresponds to the alignment given in a previous example.

---

[3] A different set of edit operations may be defined to suit a particular application. For example, in text processing, a swap of two adjacent characters may be considered an edit operation. However, a different algorithm than presented here may be required to accomodate these additional edit operations.

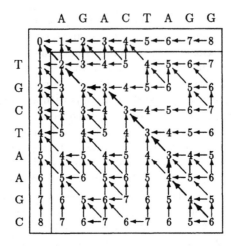

Fig. 1. Dynamic programming table with minimization pointers

2 SPLASH Reconfigurable Logic Array

SPLASH is a reconfigurable logic array developed at the Supercomputer Research Center (SRC) as a coprocessor card for the Sun VME bus [2]. As depicted in Fig. 2, the SPLASH board contains 32 Xilinx XC3090 field-programmable gate arrays (FPGA) [8] with local connections to 32 1M-bit (128K by 8) static RAM chips. The FPGA's are connected linearly in a ring with input coming from a 32-bit FIFO queue connected to chip 0 and output going to a 32-bit FIFO queue connected to chip 31. A RAM chip is connected between each pair of adjacent FPGA chips and can be accessed by either FPGA. The data path connecting the FIFO's to the array consists of 36 unidirectional lines, 32 for data and 4 for control signals. Adjacent FPGA's, except for chips 0 and 31, are joined by a 68-bit programmable bidirectional bus, which shares connections to the local RAM. Chips 0 and 31 are connected with a 35-bit data path. This "wrap-around" connection allows data flow through the array in either direction.

At the heart of the SPLASH board are the Xilinx XC3090 FPGA's. Each FPGA contains 320 configurable logic blocks (CLB's) arranged in a 20 × 16 grid and surrounded by 144 input/output blocks (IOB's). The 144 IOB's surrounding each XC3090 FPGA provides connections to the control bus and programmable interconnections between adjacent chips and local RAM. Each IOB can be configured as either an input port, an output port, or a bidirectional input/output port, with optional latch or flip-flop operation. The programmability of the IOB's allows for flexibility in the interchip connections. For example, when the local RAM is not needed, it can be disabled and the IOB's connected to the RAM's address and data lines can be used for communication between adjacent FPGA's.

The reader is referred to [2] for a more complete description of SPLASH.

Using the FPGA technology in SPLASH, we were able to rapidly prototype

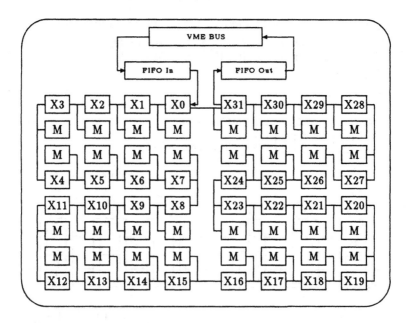

Fig. 2. The SPLASH Reconfigurable Logic Array

the systolic array without having to construct any additional hardware. This approach also has advantages over software-only simulations in that it allowed us to detect and correct race conditions present in early prototypes.

3 Systolic Array for Sequence Alignment

Our systolic array for sequence alignment operates in two phases. In the first phase, the systolic array operates in sequence comparison mode to compute entries in the dynamic programming table and store them in local RAM. In the second phase, the stored table is used to construct an alignment with a *marker passing* systolic array.

Since the data path to local RAM is 8-bits wide, for convenience, eight processing elements (PE's) were placed on each FPGA chip, except for X0 and X31, where only four PE's were placed to leave room for I/O logic. This puts a total of 248 PE's on SPLASH, allowing for alignment of sequences up to 123 in length.

3.1 Phase One: Dynamic Programming

The dynamic programming recurrence can be mapped onto a linear systolic array that computes a single antidiagonal of the dynamic programming table at each step, with each PE in the array computing the distances along one diagonal. The resulting systolic array (Fig. 3) and its implementation on SPLASH is described in [3]. The array is modified to save the dynamic programming table in local RAM. The first phase ends just after the edit distance, $d_{m,n}$, has been computed.

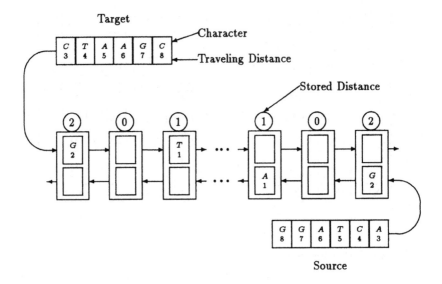

Fig. 3. Systolic array for sequence comparison

Figure 4 shows a block diagram of a sequence comparison PE. Each PE is implemented in 13 CLB's, eight for the character comparator and five for the finite state machine.

3.2 Phase Two: Marker Passing

The pointer traceback procedure for constructing an alignment, as described earlier, is performed systolically in the second phase. We can think of the traceback as a *marker passing* process in which the marker hops along a path created by the minimization pointers. Using the same antidiagonal mapping of the dynamic programming table to PE's as in phase one, we seek to move the marker from the lower-right corner of the table to the upper-left. Following a horizontal pointer would correspond to moving the marker left one PE. Similarly, following a vertical pointer corresponds to moving the marker right one PE. Finally, following a diagonal pointer corresponds to keeping the marker in the same PE. Where there are multiple pointers, one is arbitrarily chosen. In phase one, the minimization pointers were never actually computed. However, we can deduce the pointers originating from position (i, j) given $d_{i,j}$, $d_{i-1,j}$, $d_{i-1,j-1}$, s_i, and t_j. Therefore, by reading back the distances saved in local RAM and streaming the source and target sequences backwards through the array, the minimization

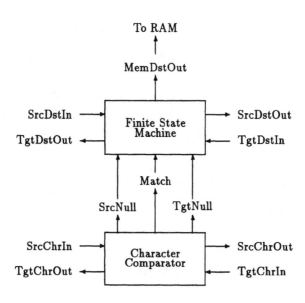

Fig. 4. Block diagram of the sequence comparison PE

pointers, and thus the movement of the marker, can be computed. The algorithm outlined above is realized by the systolic alignment array diagrammed in Fig. 5.

The sequence alignment PE is diagrammed in Fig. 6. The sequence alignment array uses the same character comparator in the sequence comparison array. The additional finite state machine is implemented in eight CLB's, bringing the total number of CLB's per PE for both phases to 21.

Since at any step, the marker can move at most one PE from its current position, the marker can be registered on a systolic stream that moves across two PE's at each step. The output of the marker stream encodes the movement of the marker. Two consecutive 1's indicate that the marker moved right. Two 0's between successive 1's indicate that the marker moved left. A pattern of 10101 indicates that the marker did not move. The binary pattern exiting the marker stream can be decoded into a series of edit operations by a simple finite state automaton that counts the number of 0's between successive 1's.

4 Benchmarks

For timing, we performed 10,000 alignments of 100-long sequences on SPLASH. It took 0.50 seconds to initialize the SPLASH array and 3.2 seconds to run the alignments. Normalizing for 100 alignments gives 0.032 seconds. For comparison, the benchmarks for 100 comparisons (without alignments) of 100-long sequences found in [3] are summarized in Table 1. We have not completed benchmarking sequence alignment on conventional computers and use these results for prelim-

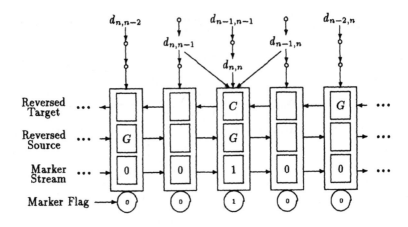

Fig. 5. Systolic array for generating alignment

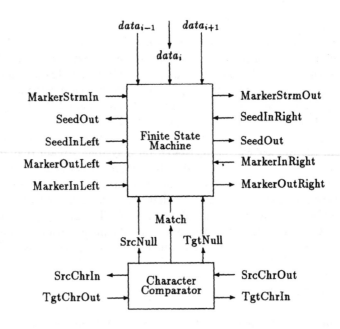

Fig. 6. Block diagram of alignment PE

inary comparison. Computing an alignment would require additional processing and therefore take additional time in most implementations. Even including initialization time, the SPLASH implementation performs at least an order of magnitude faster than implementations on commercial supercomputers, which, as tested, compute only the edit distance.

Table 1. Benchmarks of 100 comparisons of 100-long sequences [3]

System	Time	Speed-Up
SPLASH	0.020 s	2,700
P-NAC	0.91 s	60
Multiflow Trace	3.7 s	14
Sun SPARCstation 1	5.8 s	9.3
Cray 2	6.5 s	8.3
Convex C1	8.9 s	6.0
DEC VAX 8600	31 s	1.7
Sun 3/140	48 s	1.1
DEC VAX 11/785	54 s	1.0

5 Conclusion

A systolic array for sequence alignment is presented and its implementation on SPLASH is described. Preliminary benchmarks show that the SPLASH implementation is several orders of magnitude faster than implementations on supercomputers costing many times more.

References

[1] R. J. Lipton and D. P. Lopresti, "A Systolic Array for Rapid String Comparison," in *1985 Chapel Hill Conference on VLSI*, H. Fuchs, Ed. Rockville, MD: Computer Science Press, pp. 363–376, 1985.

[2] M. Gokhale, W. Holmes, A. Kopser, S. Lucas, R. Minnich, D. Sweely and D. Lopresti, "Building and Using a Highly Parallel Programmable Logic Array," *Computer*, 24, no. 1, pp. 81–89, January 1991.

[3] D. P. Lopresti, "Rapid Implementation of a Genetic Sequence Comparator Using Field-Programmable Logic Arrays," presented at Advanced Research in VLSI Conference, Santa Cruz, March 1991, Invited paper.

[4] B. A. Shapiro, "An Algorithm for Comparing Multiple RNA Secondary Structures," *Comput. Applic. Biosci.*, 4, no. 3, pp. 387–393, 1988.

[5] H. Margalit, B. A. Shapiro, A. B. Oppenheim and J. V. M. Jr., "Detection of Common Motifs in RNA Secondary Structures," *Nucleic Acids Research*, 17, no. 12, pp. 4829–4845, 1989.

[6] S. B. Needleman and C. D. Wunsch, "A General Method Applicable to the Search for Similarities in the Amino-Acid Sequence of Two Proteins," *Journal of Molecular Biology*, 48, pp. 443–453, 1970.

[7] R. A. Wagner and M. J. Fischer, "The String-to-String Correction Problem," *J.. Assn. Comput. Mach.*, 1, pp. 168–173, 1974.

[8] Xilinx, Inc., *The Programmable Gate Array Data Book*. San Jose, CA, 1991.

Using FPGAs to Prototype a Self-Timed Computer

Erik Brunvand

University of Utah, Salt Lake City, Utah 84112, USA

Abstract. The NSR (non-synchronous RISC) architecture is an architecture for a general purpose processor structured as a collection of self-timed blocks that operate concurrently and communicate over bundled data channels in the style of micropipelines [3, 6]. A 16-bit version of the NSR architecture has been implemented using Actel field programmable gate arrays (FPGAs). Each of the major components of the NSR is implemented using one or two Actel FPGA chips using a library of self-timed circuit modules [1, 2]. This prototype implementation is being used to gain experience with the NSR architecture and to gather statistics about the architectural choices. The Actel FPGAs have proven to be extremely useful in quickly prototyping this novel computer architecture.

1   Introduction

Self-timed circuits are distinguished from clocked synchronous circuits by the absence of a global synchronizing clock signal. Instead, circuit elements synchronize locally using a handshaking protocol. This protocol requires that a circuit begin operation upon receipt of a request signal and produce an acknowledge signal when its operation is complete. Self-timed circuit techniques are beginning to attract attention as designers confront problems associated with the speed and scale of modern VLSI technology [3, 6, 5]. Many of the problems associated with large VLSI systems are directly related to distributing the global clock to all parts of the system. In addition to avoiding these clock distribution problems, self-timed circuits can be faster, more robust, and easier to design than their globally clocked counterparts.

One reason for the lack of experimentation with self-timed systems is the lack of commercially available parts to support this style of design. This problem can be largely overcome by designing the special purpose parts in custom VLSI, but the turnaround time for custom chips is still too long to allow rapid prototyping and experimentation with novel systems. Field programmable gate arrays (FPGA) offer an excellent alternative for rapid development of novel system designs provided suitable circuit structures can be implemented. A library of self-timed circuit elements has been developed for use with Actel FPGAs [1, 2]. This library has been used to build a prototype of a self-timed general purpose processor.

2   NSR Architecture

The organization of the NSR can be seen in the block diagram shown in Figure 1. Each logical unit of the NSR, as shown in this figure, is its own self-timed process. The thick lines in the figure are bundled data paths and the thin lines are two-phase transition request-acknowledge style control lines [3, 6]. In addition to being internally self-timed, the units are decoupled through self-timed first-in first-out (FIFO) queues between each of the units. The logical units of the NSR are similar to standard synchronous pipeline stages such as instruction fetch, instruction decode, execute, memory interface and register file, but each operates concurrently as a separate self-timed process. The decoupling queues between each of the units allows a high degree of overlap in instruction execution. Branches, jumps, and memory accesses are also decoupled through the use of additional FIFO queues. Requesting that these types of operations execute in advance of when they are needed and that the results be queued for later use can hide the execution latency of these instructions.

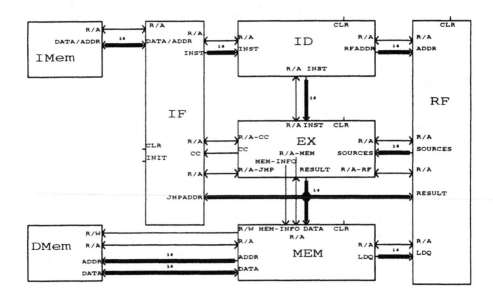

Fig. 1. NSR Architecture Block Diagram

3   NSR Instruction Set

The instruction set of the 16-bit version of the NSR is shown in Table 2. Instructions are all 16-bits wide with fixed-field decoding of opcodes and register addresses. Arithmetic and logical operations are standard three-address instructions. Other instructions are somewhat more unusual. Branch and jump instructions, for example, are decoupled through FIFO queues. A jump instruction must be preceded in the instruction stream by an instruction that computes the jump address. This instruction, SJMP, uses the ALU to compute the jump address as the sum of two registers. This address is sent to a FIFO queue connecting the execute unit and instruction fetch unit. The jump instruction, JMP, dequeues an address from the queue and uses this value to update the PC value.

Mnemonic	Encoding	Action
ADD Rd,Ra,Rb	1100 -Rd- -Ra- -Rb-	Rd ← Ra + Rb
SJMP Rd,Ra,Rb	1101 -Rd- -Ra- -Rb-	Rd, Jmp-Queue ← Ra + Rb
LDA Rd,Ra,Rb	1110 -Rd- -Ra- -Rb-	Rd, AQ(load) ← Ra + Rb
STA Rd,Ra,Rb	1111 -Rd- -Ra- -Rb-	Rd, AQ(store) ← Ra + Rb
SUB Rd,Ra,Rb	0100 -Rd- -Ra- -Rb-	Rd ← Ra - Rb
AND Rd,Ra,Rb	1000 -Rd- -Ra- -Rb-	Rd ← Ra AND Rb
OR Rd,Ra,Rb	1001 -Rd- -Ra- -Rb-	Rd ← Ra OR Rb
XOR Rd,Ra,Rb	1010 -Rd- -Ra- -Rb-	Rd ← Ra XOR Rb
XNOR Rd,Ra,Rb	1011 -Rd- -Ra- -Rb-	Rd ← Ra XNOR Rb
BCND offset	0001 —offset—	If CC-Queue, PC ← PC + offset
JMP	0000 xxxx xxxx xxxx	PC ← Jmp-Queue
MVPC Rd,offset	0111 -Rd- –offset–	Rd ← PC + offset
MVIH Rd,value	0010 -Rd- –value–	Rd.H ← value, Rd.L=0
MVIL Rd,value	0011 -Rd- –value–	Rd.L ← value, Rd.H=0
SHRA Rd,Rb	0110 -Rd- 0100 -Rb-	Rd ← shift right arithmetic Rb
SHRL Rd,Rb	0110 -Rd- 0010 -Rb-	Rd ← shift right logical Rb
SHLL Rd,Rb	0110 -Rd- 0001 -Rb-	Rd ← shift left logical Rb
SEQ Ra, Rb	0101 00xx -Ra- -Rb-	CC-Queue ← (Ra = Rb)
SNE Ra, Rb	0101 11xx -Ra- -Rb-	CC-Queue ← (Ra ≠ Rb)
SGT Ra, Rb	0101 01xx -Ra- -Rb-	CC-Queue ← (Ra > Rb)
SGE Ra, Rb	0101 10xx -Ra- -Rb-	CC-Queue ← (Ra ≥ Rb)

Fig. 2. NSR Instruction Set (16-bit Prototype)

Branch instructions are similar. Branches are PC relative with the sign extended offset encoded in the instruction, but the condition code that determines if the branch is taken must be computed in advance using a set-condition instruction. These instructions compute a condition by comparing a pair of registers and queue the resulting condition bit in a FIFO queue. The branch instruction uses this bit to decide whether to branch or not. If the condition code is computed far enough in advance of the branch instruction, there is no delay in taking the branch. This is similar to filling a branch delay slot in a synchronous pipeline. The key difference is that there is no fixed number of slots between the generation of a condition bit and the use of that bit. If there are no instructions between the setting of the condition bit and the use of that bit the NSR pipeline will simply pause to wait for the bit to be computed. If there are enough instructions, the condition bit will be waiting when the branch is executed and there will be no delay.

Load and store instructions go through FIFO queues to memory in a manner similar to the WM [8] and PIPE [4] processors. The NSR has no instructions that correspond exactly to a standard load or store instruction. Instead, instructions compute addresses for a load or a store and enter those addresses into a FIFO queue in the memory interface. Access to data from the memory is through additional queues. Data accesses to and from memory look like register accesses to the programmer. Register R1 is the special register used for these accesses. Reading from register R1 results in a value that has been fetched from memory, and writing into register R1 queues up data to be sent to memory. The memory interface has the responsibility of taking data written to R1, and addresses queued up as store addresses and writing to memory. For a load, the memory interface takes a load address from the address queue, reads from memory, and queues up the data against future reads of register R1. The layout is shown in Figure 3. The important thing to note from the instruction set in Table 2 is that the LDA (compute load address) and SDA (compute store address) instructions do not deal with data at all. They only compute addresses and queue up those addresses for use in the memory interface. References to register R1 in other instructions will provide or use the data.

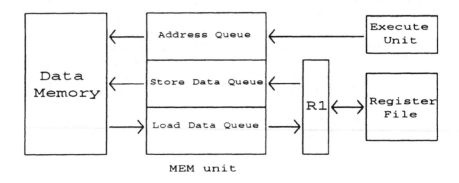

Fig. 3. NSR Memory Queues

4   FPGA Implementation

The processes that implement the separate pieces of the prototype NSR are each implemented using Actel FPGAs. These chips have been assembled as a wire-wrapped prototype for testing and evaluation. The number of Actel FPGA chips used to implement each of the parts of the NSR and the utilization of those chips are shown in Table 4. These parts were designed and implemented by students in a graduate seminar on VLSI architecture using the Workview suite of schematic capture and simulation tools from ViewLogic [7].

System Piece	Chips Used	Logic Modules	Percent Utilization
Instruction Fetch	1 Actel 1020A	547	100%
Instruction Decode	1 Actel 1010A	287	97%
Execute	1 Actel 1020A	518	95%
Register File	2 Actel 1020A	1076	98%
Memory Interface	2 Actel 1010A	554	94%

Fig. 4. NSR FPGA Implementation

The instruction fetch unit gets instructions from the instruction memory (IMem) and passes them to the instruction decode unit. It also holds the program counter (PC), and therefore processes branch and jump instructions directly. Jump addresses are generated in the execute unit and passed to the instruction fetch unit through a single-place FIFO queue. Condition codes are likewise computed in the execute unit and passed to the instruction fetch unit through a FIFO queue, this time 8 places deep.

The instruction decode unit takes the instruction from the instruction fetch unit (through a FIFO queue) and decodes information for both the register file and execute units. The register file receives register address information, and the execute unit receives decoded instruction bits.

The execute unit receives its instructions in pre-decoded form from the instruction decoder. It uses that information, and the necessary operands from the register file, to produce a result. This result can be routed back to the register file, to the memory unit, or, in the case of computing a jump address or a condition code, to the instruction fetch unit.

The register file receives register address information from the instruction decoder. It passes operands to the execute unit and receives results from the execute unit. It also receives data from the memory interface through the Load Data Queue (LDQ). Data loaded from the data memory (DMem) is available by reading a special register, R1, in the register file. When this register is accessed, the register file requests data from the memory unit. Data written to R1 is sent to the Store Data Queue (SDQ) in the memory interface to be stored in data memory.

The memory interface to the NSR contains the FIFO queues shown in Figure 3. The LDA and SDA instructions queue up addresses into an address queue (AQ), and the data to be loaded or stored is also queued up by reading or writing to a special register. When the memory interface has both an address, and, in the case of a write, the necessary data, a memory access is initiated. If the operation is a load, the data from memory is queued up and is available by reading a special register. This memory queue organization is similar to the WM machine [8], and the PIPE processor [4]. Our version is shown in Figure 3.

4.1 Implementation Challenges

The biggest challenge in constructing the NSR, in spite of the fact that self-timed circuits were used, was in specifying the timing of the individual circuit modules. Although the modules used to build the NSR are self-timed at their external interfaces, there can be timing constraints inside the modules themselves. Because these modules are designed to be implemented as small self-contained units, these internal timing constraints can be controlled when the module is designed. However, the placement and routing tools for the FPGAs do not allow macro cells to be placed as units. These tools may break larger modules up into their component parts and spread them around the FPGA which can result in timing violations in the modules themselves.

This is also a problem for the one timing constraint that is required external to the modules, namely the bundling constraint. The bundling constraint requires that the data bundle arrive at the receiver before the associated request signal. This is a simple condition to state, but is not a condition that standard timing tools understand. If a net is declared critical, the routing tool will attempt to make propagation of that signal fast, but if a net is not declared critical, the routing tool may happen make that net fast as well. There is no provision for declaring a net to be slow, or even declaring a net to be slower or faster than another specific net.

In spite of these potential problems, the completed modules do work together in the finished chips. The placement and routing tools seem to do a good job of clustering tightly connected components which minimizes the internal delays. A few internal timing violations did occur which were fixed by changing the design slightly and rerouting the chip. After a small number of iterations, the individual modules all worked correctly. A few bundling constraint problems had to be solved by adding small delays to the request lines and rerouting the chip.

Memory presented another problem. Because the processor is self-timed, it expects to send out a memory request, and receive and acknowledge when the data is available. Commercial memory chips unfortunately do not provide such an acknowledge signal. We use digital delay chips to delay the outgoing request transition long enough to account for the delay through the memory chips. This delayed request becomes the acknowledge to the NSR processor. This allows us to use standard static RAM chips in the NSR prototype.

5 Conclusions

Using a library of self-timed modules, a prototype of a self-timed general purpose processor has been constructed. This processor has many novel features and the FPGA implementation is being used to gain experience with the architecture before starting to implement a larger version of the processor in semicustom CMOS. We have found that field programmable gate arrays are an excellent medium for fast inexpensive system prototyping provided the necessary circuit primitives can be implemented. The Actel FPGA, and the self-timed library, have proven to be a very useful and flexible tool.

6   Acknowledgments

Many thanks to my students in the autumn 1991 VLSI Architecture class for designing and implementing the FPGA components that are the NSR processor. They are: IF stage - John Hurdle, Lüli Josepheson, ID - Ajay Khoche, Bill Richardson, Michael Stephenson, RF - Kent Bunker, V. Chandramouli, Dewey Jones, EX - Corby Bacco, Prabhat Jain, Cliff Miller, MEM - Madhu Penugonda, Marshall Soares. Thanks also to Nick Michell and Bill Richardson for helping define the NSR architecture.

References

1. Erik Brunvand. A cell set for self-timed design using actel FPGAs. Technical Report UUCS–91–013, University of Utah, 1991.
2. Erik Brunvand. Implementing self-timed systems with FPGAs. In W. R. Moore and W. Luk, editors, *FPGAs*, chapter 6.2, pages 312–323. Abingdon EE&CS Books, 1991.
3. Erik Brunvand. *Translating Concurrent Communicating Programs into Asynchronous Circuits*. PhD thesis, Carnegie Mellon University, 1991. Available as Technical Report CMU-CS-91-198.
4. J. R. Goodman, J Hsieh, K Liou, A. R. Pleszkun, P. B. Schechter, and H. C. Young. PIPE: A VLSI decoupled architecture. In *12th Annual International Symposium on Computer Architecture*, pages 20–27. IEEE Computer Society, June 1985.
5. Alain J. Martin. Compiling communicating processes into delay insensitive circuits. *Distributed Computing*, 1(3), 1986.
6. Ivan Sutherland. Micropipelines. *CACM*, 32(6), 1989.
7. ViewLogic Corporation. *Workview Reference Manual*, 1991.
8. Wm. A. Wulf. The WM computer architecture. *Computer Architecture News*, 16(1), March 1988.

Using FPGAs to Implement a
Reconfigurable Highly Parallel Computer

Arne Linde*, Tomas Nordström‡, and Mikael Taveniku*

* Department of Computer Engineering	‡ Division of Computer Engineering
Chalmers University of Technology,	Luleå University of Technology,
S-41296 Göteborg, Sweden	S-95187 Luleå, Sweden

E-mail: arne@ce.chalmers.se, tono@sm.luth.se, micke@ce.chalmers.se

Abstract. With the arrival of large Field Programmable Gate Arrays (FPGAs) it is possible to build an entire computer using only FPGA and memory. In this paper we share some experience from building a highly parallel computer using this concept. Even if today's FPGAs are of considerable size, each processor must be relatively simple if a highly parallel computer is to be constructed from them. Based on our experience of other parallel computers and thorough studies of the intended applications, we think it is possible to build very powerful and efficient computers using bit-serial processing elements with SIMD (Single Instruction stream, Multiple Data streams) control.

A major benefit of using FPGAs is the fact that different architectural variations can easily be tested and evaluated on real applications. In the primary application area, which is artificial neural networks, the gains of extensions like bit-serial multipliers or counters can quickly be found. A concrete implementation of a processor array, using Xilinx FPGAs, is described in this paper.

To get efficient usage and high performance with the FPGA circuits signal flow plays an important role. As the current implementation of the Xilinx EDA software does not support that design issue, the signal flow design has to be made by hand. The processing elements are simple and regular which makes it easy to implement them with the XACT Editor. This gives high performance, up to 40–50 MHz.

1 Introduction

The requirements for flexibility and adaptivity to different circumstances and environments have motivated research and development towards trainable systems rather than programmed ones. This is true especially for "action oriented systems" which interact with their environments by means of sophisticated sensors and actuators, often with a high degree of parallelism [2]. Response time requirements and the demand to accomplish the training task points to highly or massively parallel computer architectures.

In REMAP, the Real-Time, Embedded, Modular, Action-oriented, Parallel Processor Project [3], the potential of distributed SIMD (Single Instruction stream, Multiple Data streams) modules for realization of trainable systems is investigated. Each SIMD module is a highly parallel computer with simple PEs tuned to efficiently compute artificial neural network algorithms.

Within the project, a series of studies have been performed [10–12, 16] concerning the execution of neural network algorithms on highly parallel SIMD computers, with special emphasis on architectures based on bit-serial processing elements (PEs). The results show that SIMD is the best suited parallel processing paradigm for artificial neural networks (ANNs) and that arrays of bit-serial PEs with simple inter-PE communication are surprisingly efficient. As multiplication is found to be the single most important operation in these computations, there is much to be gained in the bit-serial architecture if support for fast multiplication is added.

Using today's relatively large field-programmable gate arrays (FPGAs), it is possible to build an entire computer using only FPGAs and memory. Still, if a highly parallel computer is to be constructed out of them, each processor must be very simple. As shown in our studies of parallel computers for ANN, bit-serial PEs with SIMD control suit our computational needs, which makes it feasible to use FPGAs as a means to construct the first prototypes of our computers.

The computer built should not be seen as a final "product", it is more of an architecture laboratory, in which it is possible to change the architecture of each PE rapidly. Designing and compiling a new architecture takes about one week and downloading an already prepared architecture takes less than a second.

2 Applications

To realize action-oriented systems, the artificial neural network (ANN) models [6, 7] form a very important implementation class. As shown in [12] the demands on the architecture are quite moderate for standard ANN algorithms like feed-forward networks with back propagation, Hopfield networks, or Kohonen self-organizing maps. These models, like most of the ANN algorithms, use a very simple model of the neuron. Typically, an artificial neuron computes a weighted sum of its inputs, a nonlinear function is then usually applied to the sum, and the result is sent along to neighboring neurons, as shown in Fig. 1. The power of ANN computations comes from the large number of neurons (nodes) and their rich interconnections via synapses (weights).

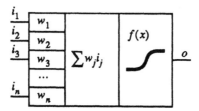

Fig. 1. The simplest model of a neuron. The neuron calculates the weighted sum of its inputs and applies a non-linear function to it, $o = f(\sum w_j i_j)$.

Different ANN models are characterized not only by the type of nodes, but also by the interconnection topology, and the training algorithm used [9]. Common topologies are layered feed-forward networks, winner take all networks, and all-to-all (Hopfield) networks. Common training rules are error back-propagation and self-organizing feature maps.

Parallelism can be found in many different places [12] but for action-oriented systems the parallelism in the nodes and weights are the important ones (node and weight parallelism). As we are focusing on the ANN models in which one can count the number of nodes and weights in thousands, we will have a lot of parallelism available. These two types of parallelism also fit the SIMD concepts perfectly.

The calculation of the weighted sum is the most time consuming calculation and should therefore be supported architecturally by any computer intended for real-time ANN computations. Also the communication means between different ANN algorithms/modules as well as between these modules and the environment have to be carefully designed.

Another possible application area for the architecture we describe would be low-level image processing. As the architecture is not very different from architectures which are known to perform well on low-level image processing problems (e.g. AIS-5000 [14], LUCAS [5]), this problem area also fits our architecture well.

3 The REMAP Computer

REMAP is an experimental project. A sequence of gradually evolved prototypes are being built, starting with a small, software configurable PE array module, implemented as a Master's thesis project [8]. With only slight modifications in the PE array architecture, but using a new high-performance control unit, the second prototype has now been built[1]. This prototype is almost full-scale with respect to the number of PEs, but far from miniaturized enough for embedded systems. It is the architecture of this prototype that is described in this paper.

The computer consists of a number of computing modules controlled by a master computer. Each computing module is a SIMD computer of its own. It contains a linear array of bit-serial processing elements with memory and I/O-circuits controlled by a control unit, as shown in Fig. 2.

---

1. A 128 PE prototype has now (beginning of 1993) been completed.

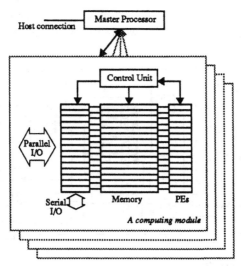

Fig. 2. Overview of the REMAP system. The PEs are implemented in Xilinx XC4005 circuits (8 in each) and the serial/parallel I/O device in Xilinx XC3020 (8 parallel and 8 serial I/O each).

3.1 The Control Unit

The main task for the control unit is to send instructions together with PE memory addresses to the PE array. At the same time it computes new address values (typically increments and decrements).

The control unit currently in use [3] has been designed around a microprogrammable sequencer and a 32 bit ALU (AMD 28331, 28332). The control unit is capable of sending out a new address together with a new instruction every 100 ns. The controller is more general purpose than usually needed, but until we know what is needed it serves our purpose. The microprograms to be executed by the control unit are stored in an 8K words control store. The operations can either be simple field operations, like adding two fields, or whole algorithms like an ANN computation. For the moment only a micro-code assembler is available to program the control unit, but we intend to develop more high level software development tools in the future. Currently we are looking into the possibility of using/developing a data-parallel language similar to C* [17].

3.2 PEs for ANN Algorithms

The detailed studies of artificial neural network computations have resulted in a proposal for a PE that is well suited for this area. The design is depicted in Fig. 3. Important features are the bit-serial multiplier and the broadcast connection. Notably, no other inter-PE connections than broadcast and nearest neighbor are needed. The PE is quite general purpose, and we are confident that this is a useful PE design also in several other application areas. In this version it consists of four flipflops (R, C, T and X), eight multiplexers, some logic and a multiplication unit. The units get their control signals directly from the micro instruction word sent from the control unit.

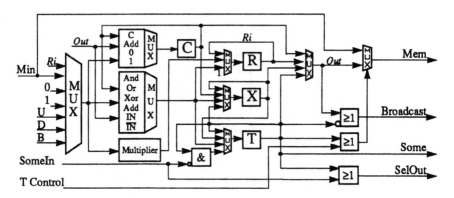

Fig. 3. The sample PE.

In simple PEs without support for multiplication the multiplication time grows quadratically with the data length. A method based on carry-save adders [5] (see Fig. 4) can reduce the multiplication time required to the time to load the operands and store the result.

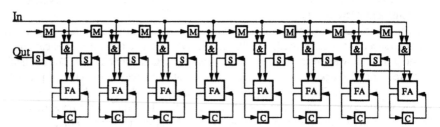

Fig. 4. Design of a two's-complement bit-serial multiplier. It is operated by first shifting in the multiplicand, most significant bit first, into the array of M flip-flops. The bits of the multiplier are then successively applied to the input, least significant bit first. The product bits appear at the output with least significant bit first.

As shown in [11] the incorporation of a counter instead of a multiplier in the PE design may pay off well when implementing the Sparse Distributed Memory (SDM) neural network model. A 256 PE REMAP realization with counters is found to run SDM at speeds 10–30 times that of an 8 K PE Connection Machine CM-2, (with frequencies normalized and on an 8 K problem). Already without counters (then the PEs become extremely simple) a 256 PE REMAP outperforms a 32 times larger CM-2 by a factor of 4–10. Even if this speed-up for REMAP can be partly explained by the more advanced sequencer, the possibility to tune the PEs for this application is equally important.

3.3 PE Communication

The processing element has two ways of communicating with other processing elements: nearest neighbor and broadcast communication. The nearest neighbor communication network allows each PE to read its neighbor's memory i.e. PE(n) can read from PE(n+1) and PE(n-1). The first and the last PEs are considered neighbors. At any time one of the PEs can broadcast a value to all other PEs or to the control unit. The control unit can also broadcast a value to the PEs. It has also a possibility to check if any of the PEs has the activity bit (T-flip-flop) set. If several PEs are active at the same time and the control unit wants one PE to broadcast, the control unit simply does a select-first operation, which selects the first active PE and deselects the rest. These communication and arbitration operations can be used to efficiently perform matrix computations as well as search and test operations sufficient for many application areas, especially artificial neural networks. To be useful in real-time applications which include interacting with a changing environment, high demands are put on the I/O-system. To meet these demands the processor array is equipped with two I/O-channels, one for 8-bit wide communication and the other for array-wide communications. This interface has a capability to run at speeds up to 80 MHz (burst) which, for a 256 PE array, implies a maximum transfer rate of 20 Gbit/s. Due to limitations in the control unit the I/O-interface currently runs at 10 MHz which reduces the transfer rate to 2.5 Gbit/s.

4 Designing with FPGA Circuits

After a market survey we found that FPGAs from Xilinx [22] would serve our needs best. The structure of the Xilinx circuits is shown in Fig. 5. The chip consists of a number of configurable logic blocks (CLB), some input-output blocks (IOB) and an interconnection network (ICN). These circuits are user programmable, thus enabling the CLB, IOB and ICN to be programmed by the user. The configuration of the on-chip configuration RAM is carried out at power up or by a reprogramming sequence. The RAM can be loaded from an external memory or from a microprocessor, the latter is used for REMAP. It takes about 400 ms to reprogram the circuits, thus enabling the master-processor to change the architecture of the processing elements dynamically during the execution of programs.

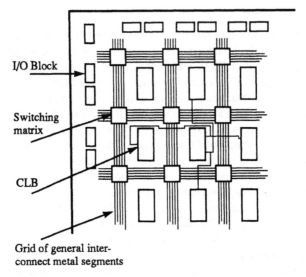

I/O Block

Switching
matrix

CLB

Grid of general inter-
connect metal segments

Fig. 5. Xilinx FPGA overview. The IOB connects the I/O-pads to the ICN. These blocks can be configured as input, output or bidirectional blocks. The CLBs are configurable logic blocks consisting of two 16bit (and one 8bit) look-up table for logic functions and two flipflops for state storage. These blocks are only connected to the ICN. The ICN connects the different blocks in the chip. It consist of four kinds of connections: short-range connections between neighboring blocks, medium-range connections connecting blocks on slightly larger distances, long-lines connecting whole rows and columns, and global nets for clock and reset signals broadcasted throughout the whole chip.

Since one of our goals is to make a kind of hardware simulator for different types of PE-architectures using a fixed hardware surrounding, it is required that the connections off chip like those to the memory and control unit have the same function regardless of the currently loaded processor architecture. As shown in [18] it is advantageous to lock the pads so that control signals enter from the top and bottom of the chip, and also design the processing elements so that they are laid out rowwise in the array of CLBs. It is likewise preferable to have a dataflow from left to right in the chip i.e. input data enters the left side and output emerges from the right side.

4.1   Using XC3090

The first prototype was constructed using Xilinx XC3090, and some frustrating experiences were gained from the poor development tools for these circuits. The processing elements in this version are only capable of running at 5MHz clock frequency. The low speed is due to the incapability of the EDA software to handle signal flow layout in the circuits, something which also leads to low utilization. The PEs were designed using the OrCAD CAE-tools, enabling the designer to work with ordinary logic blocks like multiplexers and different types of gates. The schematics are then automatically converted to suit the Xilinx circuits. This is a fast design method but different parts of the logic become intermixed and long delays are introduced.

4.2 Using XC4005

The current prototype is based on the XC4000 FPGA family from Xilinx. These circuits have a more balanced performance than the XC3000 circuits which have small routing resources compared to the number of CLBs. In the XC4000 family the CLBs are larger, the ICN much more powerful and the internal delays shorter. The circuits range from XC4003, which has a 10 by 10 CLB matrix, to XC4010 with a 20 by 20 CLB matrix, and even larger circuits are announced. With these circuits it is easier to test new types of PEs, as there is more space in them. It will also be possible to increase the maximum clock frequency to 20MHz, and possibly even 40MHz if more pipelining is introduced. The greatest advantage with these new circuits is the software; routing a XC3090 chip can take a couple of days on a 80486 machine, while the same problem can be solved in half an hour with the new software for the XC4000 circuits.

One PE of the kind depicted in Fig. 3 occupies approximately 10 CLBs and the eight-bit deep bit-serial multiplier 11 CLBs. Using a XC4005 with a 14 by 14 CLB-matrix, we can get at least 8 PEs in each Xilinx chip. Considering this and the timing demands of 10MHz operation (due to the control unit), we can easily make design variations both in the main processor and in the multiplier (or other coprocessor). It takes about one week to make a tested and simulated prototype with the XACT editor. The design is of course also open for changes to PEs with other data widths between two bits and eight bits.

Tools

High level tools were not available when we started to develop a processing element for the XC4005-circuits. Therefore we have not yet tested how well those tools work. There are, however, several advantages of using the low-level XACT editor in early stages of the design. We get good knowledge of the circuit's limitations and possibilities, and at the same time we get full control of all necessary timing. The usage of XACT is simplified by the regular and simple structure of our design. In the first implementation we aimed towards eight processing elements running at 10MHz in each XC4005 circuit, based on the previous experiences with the XC3090 circuits. These goals were easily achieved; the eight processing elements can run at 20MHz utilizing 75% of the XC4005 configurable logic blocks and all of its I/O blocks, this in the 84 pin PLCC package.

Data and Control Flow in the Circuits

The data and control flow play an important role in getting the best performance out of the circuits, therefore we have a basic template with some of the control and data signals already laid out. This template enables the user to easily implement new types of processing elements with minimum effort and at the same time achieve high performance.

When designing the control flow we want to use the global networks as much as possible. This is achieved by using 4 of the global nets and 20 of the vertical long-lines. The memory input signal is connected to the memory output via a horizontal long-line through the chip in order to enable good data input distribution and allow write-back of unchanged data when the processing element is inactive. With these restrictions in signal flow the internal delays can be held very low.

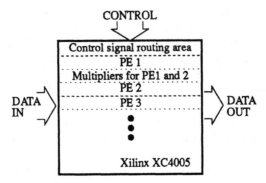

Fig. 6. Layout for PEs in an XC4005.

As we have used the XC4005-PC84 which has a 14x14 CLB matrix, and the rest of the hardware is designed for eight processing elements, the chip is divided into four blocks of two processing elements each, occupying three rows in the matrix. Each processing element then gets 21 CLBs, 2 IOBs with PADs, and six IOBs with only edge decoders[1]. After this we have 28 unused CLBs.

Testability

From our experiences with the XC3090 circuits, which sometimes got into undefined states when we tried to reconfigure them, we now separate programming pins to the master processor so that we can directly see which circuit is failing. We also use the possibility of reading back configuration and state data from the Xilinx circuits, which can be done while the PEs are running. The master can also single-step the processor via the control unit and read back all state variables. Two pads on each processing element are dedicated to probing, here we can measure any internal delay simply by loading a configuration with the probe outputs properly programmed (this is done automatically by the XACT EDA software). The JTAG facilities of the XC4000 have not been used, because the PEs, simple as they are, only require a couple of hundred stimuli to excite all modes in them.

The full-scale prototype (256 PEs) can run in 10 MHz with very comfortable timing margins. More memory and additional communication networks can easily be added if need arises.

---

1. Some of the I/O-blocks in the XC4005-PC84 have no connections to pads. However, these blocks can be used to get a connection to their edge decoder.

5 Other Usage of FPGAs to Run ANN Algorithms

There are other FPGA implementations of ANN models besides ours. A short description of some of them are given below.

A group at North Carolina State University has developed a PC-card called Anyboard [19], which in principle only contains Xilinx chips (4 XC3020s) and RAM. It is part of a "rapid prototyping" environment, where user-specified digital designs can quickly be implemented and tested. One early project using this card was the implementation of a stochastic ANN model called TInMANN [20]. A quite fast and dense implementation was obtained. They used a special purpose architecture, tuned to their algorithm.

Another project, using 25 XC3020s to implement a stochastic Boltzmann machine ANN, was carried out by Skubiszewski [15]. In this implementation the architecture was more like ours (identifiable PEs similar to conventional bit-serial PEs), but no support for the multiplications was included.

Cox and Blanz [4] built an ANN simulator with impressive performance, out of 28 XC3090s. In contrast to the two implementations above and our implementation, they have used a highly specialized bit-parallel approach, which implements a feed-forward neural network of a fixed size (12x14x4).

Another, more specialized, use of FPGAs for ANN computations is made by a group at Tampere University of Technology, Finland [13]. In this group's hardware implementation of Kanerva's Sparse Distributed Memory (SDM), FPGAs are used to implement the main controller as well as more specialized computations like an adder tree. The architecture is highly specialized for SDM and no identifiable processing elements exist.

Xilinx circuits are also used in general hardware emulators such as the Quickturn RPM emulator [21], which emulates designs with from 10K up to 1M gates at a speed of 1MHz. This type of emulators could of course be used to simulate all the designs described in this paper, but with drastically lower speed and CLB utilization.

6 Conclusions and Future Directions

With the REMAP computer, we have a platform from where we can test and evaluate different types of interconnection networks, PE complexities and architectures. This is not restricted to simple bit-serial PEs as the one described in this text, also complex ones such as bit-serial floating point arithmetic units and up to eight bit wide PEs can be implemented. Floating point arithmetic for this platform has been examined by members of the group [1], and will be included. When we have found a good PE architecture we will transfer it to silicon, this decreases the size and increases the system speed. Our aim is to get 256 processing elements, with floating point arithmetic, on each chip running at an internal speed of 200–300MHz.

A robot arm with 12 motors and a number of sensors all controlled in parallel from the array-parallel interface on the REMAP computer is being developed at the Centre for Computer Architecture, Halmstad University. A CCD camera is also planned to be connected to the byte-wide interface on REMAP as a further step towards a real-time action-oriented system.

To speed up the development cycle in the future some sort of high-level description of the PEs and their interconnections would be needed. From this description it should be possible to generate FPGA layout, VLSI layout, a PE array simulator, and a high level language compiler back-end. Both text-based and graphics based high-level descriptions are considered.

While, in this design of a hardware simulator, we are more interested in the possibilities of changing the processor architecture than to get maximum performance, we have added (retained) the feature that design changes can be made during execution. For example in some parts of an application we may need a counter instead of a multiplier. It is easily accomplished, via program control, to stop the control unit during approximately 400ms and reprogram the Xilinx circuits.

7 References

1. Åhlander, A. and B. Svensson. "Floating point calculations in bit-serial SIMD computers." In *Fourth Swedish Workshop on Computer Systems Architecture*, Linköping, Sweden, 1992.

2. Arbib, M. A. "Schemas and neural network for sixth generation computing." *Journal of Parallel and Distributed Computing*. Vol. 6(2): pp. 185-216, 1989.

3. Bengtsson, L., A. Linde, T. Nordström, B. Svensson, M. Taveniku and A. Åhlander. "Design and implementation of the REMAP³ software reconfigurable SIMD parallel computer." In *Fourth Swedish Workshop on Computer Systems Architecture*, Linköping, Sweden, 1992.

4. Cox, C. E. and W. E. Blanz. "GANGLION — A fast field programmable gate array implementation of a connectionist classfier." (RJ 8290 /75651/), IBM Research Division, Almaden Research Centre, 1990.

5. Fernström, C., I. Kruzela and B. Svensson. *LUCAS Associative Array Processor - Design, Programming and Application Studies*. Vol 216 of *Lecture Notes in Computer Science*. Springer Verlag. Berlin. 1986.

6. Hertz, J., A. Krogh and R. G. Palmer. *Introduction to the Theory of Neural Computations*. Addison Wesley. Redwood City, CA. 1991.

7. Kohonen, T. "An introduction to neural computing." *Neural Networks*. Vol. 1: pp. 3-16, 1988.

8. Linde, A. and M. Taveniku. "LUPUS — a reconfigurable prototype for a modular massively parallel SIMD computing system." (Masters Thesis 1991:028 E), University of Luleå, Sweden, 1991. [In Swedish]

9. Lippmann, R. P. "An Introduction to Computing with Neural Nets." *IEEE Acoustics, Speech, and Signal Processing Magazine*. Vol. 4(April): pp. 4-22, 1987.

10. Nordström, T. "Designing parallel computers for self organizing maps." (Res. Rep. TULEA 1991:17), Luleå University of Technology, Sweden, 1991.

11. Nordström, T. "Sparse distributed memory simulation on REMAP3." (Res. Rep. TULEA 1991:16), Luleå University of Technology, Sweden, 1991.

12. Nordström, T. and B. Svensson. "Using and designing massively parallel computers for artificial neural networks." *Journal of Parallel and Distributed Computing.* Vol. 14(3): pp. 260-285, 1992.

13. Saarinen, J., M. Lindell, P. Kotilainen, J. Tomberg, P. Kanerva and K. Kaski. "Highly parallel hardware implementation of sparse distributed memory." In *International Conference on Artificial Neural Networks*, Vol. 1, pp. 673-678, Helsinki, Finland, 1991.

14. Schmitt, R. S. and S. S. Wilson. "The AIS-5000 parallel processor." *IEEE Transaction on Pattern Analysis and Machine Intelligence.* Vol. 10(3): pp. 320-330, 1988.

15. Skubiszewski, M. "A hardware emulator for binary neural networks." In *International Neural Network Conference*, Vol. 2, pp. 555-558, Paris, 1990.

16. Svensson, B. and T. Nordström. "Execution of neural network algorithms on an array of bit-serial processors." In *10th International Conference on Pattern Recognition, Computer Architectures for Vision and Pattern Recognition*, Vol. II, pp. 501-505, Atlantic City, New Jersey, USA, 1990.

17. Thinking Machines Corporation. "C* User's guide and C* Programming Guide." (Version 6.0), T M C Cambridge, Massachusetts, 1990.

18. Unnebäck, M. "Gate array implementations of processing elements for a reconfigurable, modular, massively parallel SIMD computer." (Masters Thesis 1991:117 E), Luleå University of Technology, 1991. [In Swedish]

19. Van den Bout, D. E., J. N. Morris, D. Thomae, S. Labrozzi, S. Wingo and D. Hallman. "AnyBoard: An FPGA-based, reconfigurable system." *IEEE Design & Test of Computers.* (September): pp. 21-30, 1992.

20. Van den Bout, D. E., W. Snyder and T. K. Miller III. "Rapid prototyping for neural networks." *Advanced Neural Computers.* Eckmiller ed. North-Holland. Amsterdam. 1990.

21. Wolff, H. "How Quickturn is filling a gap." *Electronics.* (April): 1990.

22. XILINX. *The Programmable Gate Array Data Book.* 1990.

Novel High Performance Machine Paradigms and Fast-Turnaround ASIC Design Methods: a Consequence of, and, a Challenge to, Field-programmable Logic

A. Ast, R. Hartenstein, R. Kress, H. Reinig, K. Schmidt

Department of Computer Science, University of Kaiserslautern, W-6750 Kaiserslautern, Germany

Abstract. New high performance computational paradigms have been introduced, such as Xputers. Xputers have a reconfigurable ALU using FPGA-like technology. This results in an efficient novel machine paradigm, competitive to many ASIC solutions. It permits systematic derivation of machine code from high level algorithm specs or programs. After testing and debugging real gate array specs may be derived by retargeting. This is a shortcut on the way from algorithm to silicon: less effort and shorter time to market. Compared to conventional ASIC design this means: a) real execution instead of simulation, b) higher source language level and thus more concise specification.

1 Introduction

An increasing number of researchers uses field-programmable logic for more sophisticated concepts, than just ASIC prototyping: ASIC emulation [1, 2], simulation acceleration [3], customized (von Neumann) computers [4], a universal smart memory methodology [5]. This paper deals with FPL-based implementation concepts for Xputers. The Xputer paradigm is a high performance alternative to the von Neumann machine paradigm. All this work indicates an emerging completely new branch of computing science, where fundamental procedural models of universal computational paradigms are no more based on sequential program code, being scanned from a memory, but on structural programming [6].

Xputers have a reconfigurable ALU (called rALU), based on FPGA-like technology. Fig. 1 shows the basic structure of the rALU for the MoM-4 Xputer architecture [7]. This rALU includes a number of high performance operator resources (fig. 1: h-functors, where h stands for hardwired, such as ALUs, Multipliers etc.), a field-programmable logic part for residual control [8] and customized operators (f-functors, where f stands for field-programmable), as well as four scan caches (scan windows to scan primary memory space: size and dimension are adjustable at run time [9], currently up to 5 by 5 words, or 2 dimensions, respectively).

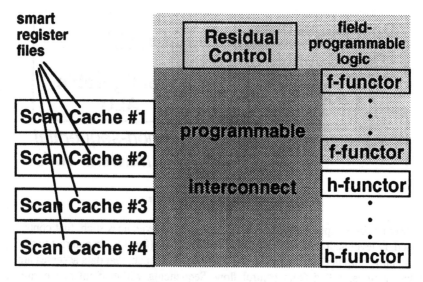

Fig. 1. rALU of the MoM-4

2 Reconfigurable Super ALU (rALU) for Acceleration

Such a rALU supports the configuration of one ore more powerful compound operators, which provide some intra-ALU parallelism (fig. 4 a) reducing the need to store and load intermediate values. A global interconnect-programmable part (see fig. 1) gives the flexibility needed to connect problem-oriented compound operators to the scan caches. In contrast to microprogrammable (e. g. von Neumann) ALUs, such compound operators are set up at loading time to avoid massive multiplexing overhead, addressing overhead, and control overhead [7, 9, 10]. The basic ideas for using such rALUs efficiently have been published elsewhere (such as: execution mechanisms [10], data-procedural programming languages and programming techniques [11, 12], compilation techniques [13], acceleration of applications in areas, such as image preprocessing [14], computer graphics [15], pattern recognition [16], digital signal processing [17], neural network emulation [18], VLSI design automation [19], and other areas [20]).

3 New Machine Paradigms Required for Structural Programming

The consequence of this is, that an Xputer does not have a hardwired instruction set, and, that an essential part of the machine code is combinational (fig. 2). That's why the rALU does not support any instruction sequencing (such as known from the von Neumann machine paradigm), so that another sequencing method is needed. That's why for Xputers data sequencing is used with a data counter (see fig. 2) - instead of the program counter known from von Neumann. Thus this creates the need for a fundamentally new basic computational model, i. e. for a completely new branch of computer science (see introduction). Figures 3 a and b illustrate the fundamental difference of

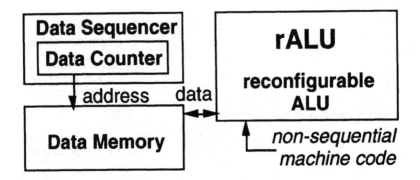

Fig. 2. Xputer Data Counter

the driving force: xputers are driven by data address sequences such, that control is a secondary level action to be evoked only upon request (fig. 3 b), in contrast to the (von Neumann) computer, driven by control flow (fig. 3 a), such, that data accesses are secondary actions, called from control flow. Sequencing within Xputers we call data sequencing (to distinguish this from the principles of so-called data flow machines, which in fact are driven by arbitration, or, by firing). The data-procedural implementation of an algorithm on an Xputer we sometimes call a data schedule for clear distinction from the familiar control-procedural von-Neumann-based implementations called programs.

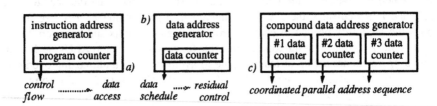

Fig. 3. Xputers vs. von-Neumann

4 The New Paradigm is much more Efficient

The data-procedural paradigm of Xputers efficiently supports much more efficient compilation techniques, than possible for (von Neumann) computers. The derivation of data sequences by data dependency analysis from high level algorithm specifications is more direct than generating control-driven conventional machine code. The machine code obtained by the new method is shorter and much less overhead-prone than von Neumann type machine code for a number of reasons explained elsewhere [19, 20]. That's why by this method highly efficient implementations can be achieved· on a simple and cheap hardware. Often such solutions are competitive to ASIC solutions, although being based on a sequential paradigm. Acceleration factors (benchmark comparisons) by mostly two and up to three orders of magnitude have been obtained experimentally for several example algorithm implementations on a monoprocessor MoM Xputer [22] (compared to technologically comparable von Neumann processors). These experiences indicate the feasibility of the following Xputer application scenarios: Xputers as universal accelerator co-processors on extension boards within workstations, and, the Xputer paradigm as basis of a more efficient new ASIC design method.

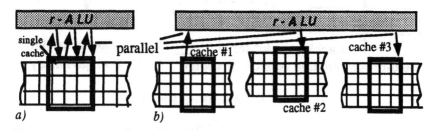

Fig. 4. Fine-Grain Parallelism Within the rALU

5 A Fast Turn-Around ASIC Design Method

This new ASIC design method is summarized by fig. 5, which is derived from am experimental environment having been implemented at Kaiserslautern [11]. An optimizer (a kind of program generator) accepts a high level specification and generates a program expressed in a high level programming language, from which the compiler generates "machine code" for a programmable version of an Xputer. Debugging turn-around is very fast, since real execution is used (instead of the highly inefficient simulation needed in other design environments). As soon as debugging is completed, a retargeting software will convert the machine code into a real (non-field-programmable) gate array design. Thus by retargeting a hard-coded hardwired version of the Xputer application is created. The compilation technique thus obtained is more than just silicon compilation. It starts with a more concise problem description input at a higher source language level, and, it generates target hardware, programs, and machine code at the same time, whereas silicon compilers only generate the target hardware. This new technique is also a new direction in high level synthesis.

6 The Super rALU is the Challenge

The extraordinarily good efficiency of the new paradigm and the surprisingly good performance figures obtained with it are highly promising. A challenge, however, because being a critical issue is the underlying field-programmable hardware technology. Especially the use of more than one scan cache [19] (fig. 3 c) in algorithms like FFT [10] creates a kind of parallelism (compare fig 4 b), where several register files communicate with each other through a common compound operator (on a kind of super rALU). For such an approach field-programmable circuits currently available commercially tend to yield extremely area-inefficient solutions. They mainly support glue logic solutions. That's why we need completely new approaches to the micro-

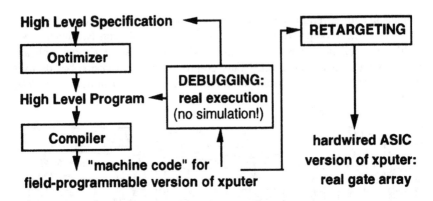

Fig. 5. New ASIC Design Method

architecture of field-programmable circuits. We need several different new micro-architectures of interconnect-reprogrammable ALUs, specialized for several application areas (digital signal processing, image processing, mathematics, high performance number crunching etc.). For some applications single-chip rALU solutions are feasible, such as e. g. for pattern matching [9, 14], where we designed our own field-programmable circuit (DPLA = Dynamically Programmable Logic Array, see fig. 6), since those available commercially did not meet out requirements. But for a high performance universal super rALU currently a MCM implementation would be needed.

Fig. 6. A DPLA IC on the MoM rALU Board

References

1. P. A. Kaufmann: Wanted: Tools for Validation, Iteration; Computer Design, December 1989

2. M. D'Amour, et al.: ASIC Emulation cuts Design Risc; High Performance Systems, October 1989

3. L. Lindh, K. D. Müller-Glaser, H. Rauch: A Real Time Kernel - Rapid Prototyping With VHDL and FPGAs; (submitted for this workshop)

4. I. Buchanan, T. A. Kean: The Use of FPGAs in a Novel Computing Subsystem; Proc. 1st Int'l ACM/SIGDA Workshop on Field-programmable Logic, Berkeley, CA, 1992

5. P. Bertin, D. Roncin, J. Vuillemin: Introduction to Programmable Active Memories; Proc. 3rd Int'l Conf. on Systolic Arrays, Kilarney, Ireland, May 1989.

6. J. P. Gray, T. A. Kean: Configurable Hardware: A New Paradigm for Computation; in: (ed.) C. L. Seitz: Advanced Research in VLSI; MIT Press, 1989

7. R.W. Hartenstein, A.G. Hirschbiel, M.Weber: A Novel Paradigm of Parallel Computation and its Use to Implement Simple High Performance Hardware; CONPAR '90 - VAPP IV, Zürich, Schweiz, Sept. 1990.

8. R. W. Hartenstein, A. G. Hirschbiel, M. Riedmüller, K. Schmidt, M. Weber: A High Performance Machine Paradigm Based on Auto-Sequencing Data Memory; HICSS-24 Hawaii International Conference on System Sciences, Poipu, Koloa, Hawaii, USA, January 1991

9. R. W. Hartenstein, A. G. Hirschbiel, M. Weber: MoM - Map Oriented Machine; in: Ambler et al.: (Preprints Int'l Workshop on) Hardware Accelerators, Oxford 1987, Adam Hilger, Bristol 1988

10. R. Hartenstein, A.G. Hirschbiel, M.Weber: The Machine Paradigm of Xputers and its Application to Digital Signal Processing Acceleration; 1990 Int'l Conf. on Parallel Processing, St. Charles, Ill, USA, Aug 1990.

11. M. Weber: An Application Development Method for Xputers; Ph. D. dissertation, Fachbereich für Informatik, Universität Kaiserslautern, December 1990

12. A. G. Hirschbiel: A Novel Processor Architecture based on Auto Data Sequencing and Low Level Parallelism; Ph. D. dissertation, Fachbereich für Informatik, Universität Kaiserslautern, 1991

13. R. W. Hartenstein, K. Schmidt, H. Reinig, M. Weber: A Novel Compilation Technique for a Machine Paradigm Based on Field-Programmable Logic; in Will Moore, Wayne Luk (ed.): FPGAs; Abingdon EE&CS Books, Abingdon, 1991; revised reprint from: Proc. Int'l Workshop on Field Programmable Logic and Applications, Oxford, UK 1991

14. R. W. Hartenstein, A. G. Hirschbiel, M. Weber: MoM - Map Oriented Machine, in: Chiricozzi, D'Amico: Parallel Processing and Applications, North Holland, Amsterdam / New York 1988.

15. R. Hartenstein, A. Hirschbiel, K. Lemmert, M. Riedmüller, K. Schmidt, M. Weber: Xputer Use in Image Processing and Digital Signal Proccessing; SPIE (Soc. of Photo-optical Instrumentation Engineers) Conf. on Visual Communication and Image Processing, Lausanne, Switzerland, 1990

16. R.W. Hartenstein, A.G. Hirschbiel, M. Riedmüller, K. Schmidt, M.Weber: Automatic Synthesis of Cheap Hardware Accelerators for Signal Processing and Image Preprocessing; 12. DAGM-Symposium Mustererkennung, Oberkochen-Aalen, September 1990.

17. R.W. Hartenstein, A.G. Hirschbiel, M.Weber: The Machine Paradigm of Xputers and its Application to Digital Signal Processing Acceleration; in: Deprettre (ed.): Algorithms and Parallel Architecture; North Holland, Amsterdam, 1991

18. R.W. Hartenstein, A.G. Hirschbiel, M.Weber: Using Xputers as Universal Accelerators for Neuro Network Simulation and its Applications; Int'l Neural Network Conference, INNC 90, Paris, France, July 1990.

19. R. Hartenstein, A. Hirschbiel, H. Riedmüller, K. Schmidt, M. Weber: A Novel Paradigm of Parallel Computation and its Use to Implement Simple High Performance Hardware; Info-Japan (International Conference on Information Technology), Tokyo, Japan, Oct. 1990

20. R. W. Hartenstein, H. Reinig, M. Riedmüller, K. Schmidt: A Novel Computational Paradigm: Much More Efficient Than Von Neumann Principles; 13th IMACS World Congress, Dublin Ireland, July 1991

21. R.W. Hartenstein, A.G. Hirschbiel, M.Weber: Xputers - An Open Family of Non von Neumann Architectures; Proc. of 11th ITG/GI-Conference: Architektur von Rechensystemen, München, März 1990, VDE-Verlag, Berlin, 1990.

22. R. Hartenstein, A. Hirschbiel, M. Weber: MoM - a partly Custom-Designed Architecture compared to Standard Hardware; Proc. IEEE Comp Euro '89, Hamburg, FRG, IEEE Press, Washington, DC, 1989

23. R. W. Hartenstein, A. G. Hirschbiel, M. Riedmüller, K. Schmidt, M. Weber: A Novel ASIC Design Approach Based on a New Machine Paradigm; IEEE Journal of Solid-State Circuits, Vol. 26, No. 7, pp. 975-989, July 1991; revised reprint from: Proc. ESSCIRC - European Solid-State Circuits Conference '90, Grenoble, Frankreich, September 1990

24. A. Ast, et al.: Using Xputers as Inexpensive Universal Accelerators in Digital Signal Processing; BILCON Int'l Conf. on New Trends in Signal Processing, Communication and Control, Ankara, Turkey, July 1990, North Holland, Amsterdam 1990

Author Index

Lecture Notes in Computer Science

For information about Vols. 1–629
please contact your bookseller or Springer-Verlag